T0345194

Machine Learning for Neuroscience

This book addresses the growing need for machine learning and data mining in neuroscience. The book offers a basic overview of the neuroscience, machine learning and the required math and programming necessary to develop reliable working models. The material is presented in an easy-to-follow, user-friendly manner and is replete with fully working machine learning code.

Machine Learning for Neuroscience: A Systematic Approach tackles the needs of neuroscience researchers and practitioners that have very little training relevant to machine learning. The first section of the book provides an overview of necessary topics in order to delve into machine learning, including basic linear algebra and Python programming. The second section provides an overview of neuroscience and is directed to the computer science-oriented readers. The section covers neuroanatomy and physiology, cellular neuroscience, neurological disorders and computational neuroscience. The third section of the book then delves into how to apply machine learning and data mining to neuroscience, and provides coverage of artificial neural networks (ANN), clustering and anomaly detection. The book contains fully working code examples with downloadable working code. It also contains lab assignments and quizzes, making it appropriate for use as a textbook.

The primary audience is neuroscience researchers who need to delve into machine learning, programmers assigned neuroscience related machine learning projects and students studying methods in computational neuroscience.

Machine Learning for Neuroscience
A Systematic Approach

Chuck Easttom

CRC Press
Taylor & Francis Group
Boca Raton London New York

CRC Press is an imprint of the
Taylor & Francis Group, an **informa** business

First edition published 2024
by CRC Press
6000 Broken Sound Parkway NW, Suite 300, Boca Raton, FL 33487–2742

and by CRC Press
4 Park Square, Milton Park, Abingdon, Oxon, OX14 4RN

CRC Press is an imprint of Taylor & Francis Group, LLC

© 2024 Chuck Easttom

ISBN: 9781032136721 (hbk)
ISBN: 9781032137278 (pbk)
ISBN: 9781003230588 (ebk)

DOI: 10.1201/9781003230588

Typeset in Times
by Apex CoVantage, LLC

Contents

SECTION I   Required Math and Programming

SECTION II   Required Neuroscience

SECTION III   Machine Learning

Preface

This book is quite special to me. It combines two passions of mine: neuroscience and machine learning. As you are, no doubt, aware, there are numerous books on machine learning, some directed towards data mining, some to finance, some to cyber security, etc. However, there is very little literature specifically on machine learning for neuroscience. That is the reason behind writing this book.

One goal of this book is to be accessible to the widest possible audience. That is why the three sections are designed as they are. Section I provides foundational material describing linear algebra, statistics and Python programming. For readers that don't have that skillset, or need a refresher, this section (chapters 1 through 4) provides that. However, other readers may not need this section. Section II is for those readers who may not have any neuroscience background: for example, a computer programmer who has been tasked with machine learning for neuroscience, but lacks any real knowledge of neuroscience. Now, if you are a neuroscientist or medical doctor, you will find section II (chapters 5 through 8) rather rudimentary. The goal of those chapters is to provide the interested reader with the necessary, foundational knowledge of neuroscience.

Finally, we come to section III. In chapters 9 through 14 we explore machine learning. In each of those chapters you will find fully functioning code. And every script in this book was actually, personally, tested by myself. So, I am certain they all work as written. This will allow you to ensure you can make machine learning scripts work, including several specifically for neuroscience. However, simply typing in others scripts is not really the goal of a book like this. The goal is for you to be able to create your own. So, in the later chapters, there will be lab exercises for the more adventuresome reader to try their skills. In those labs you may be given part of a script, or merely suggestions on how to write the script. You will have to develop it yourself. However, those only come after you have been presented with dozens of fully functioning scripts.

This brings us to the topic of how best to read this book. Clearly, the three sections were made to be independent. You can almost think of them as mini books within this book. You can skip section I or II if you don't need that material. But for section III, you should spend some time in each chapter. Don't rush through it. First, make all examples in that chapter work. Then perhaps experiment with changing a few parameters of the script and noting what happens. The idea is for you to finish this book quite comfortable with machine learning for neuroscience.

About the Author

Dr. Chuck Easttom is the author of 37 books, on topics such as cryptography, quantum computing, programming, cyber security and more. He is also an inventor with 25 patents and the author of over 70 research papers. He holds a Doctor of Science in cyber security, a Ph.D. in Nanotechnology, a Ph.D. in computer science, and three master's degrees (one in applied computer science, one in education and one in systems engineering). He is a senior member of both the IEEE and the ACM. He is also a distinguished speaker of the ACM and a distinguished visitor of the IEEE. Dr. Easttom is currently an adjunct professor for Georgetown University and for Vanderbilt University. Dr. Easttom is active in IEEE (Institute for Electrical and Electronic Engineers) standards groups, including several related to the topics in this book, such as:

Chair of IEEE P3123 Standard for Artificial Intelligence and Machine Learning (AI/ML) Terminology and Data Formats.

Member of the IEEE Engineering in Medicine and Biology Standards Committee. Standard for a Unified Terminology for Brain-Computer Interfaces P2731 from 2020 to present.

You can find out more about the author at his website www.ChuckEasttom.com.

Section I

Required Math and Programming

1 Fundamental Concepts of Linear Algebra for Machine Learning

INTRODUCTION

Before exploring what linear algebra is, it may be helpful to explain why you need to know it for machine learning. Linear algebra is used in many different fields including machine learning. One can certainly download a machine learning script from the internet and execute it without understanding linear algebra. However, to really delve into machine learning algorithms, one will need a basic understanding of linear algebra. If you wish to eventually move beyond the simple copying and modifying of scripts, and to truly work with machine learning, in any context, you will need a basic understanding of linear algebra. Furthermore, this entire text is directed towards machine learning for neuroscience. Neuroscience, by its very nature, demands a certain rigor.

You will discover in later chapters that it is common for data to be imported into a machine learning algorithm as a vector. Images are often loaded as matrices. These two facts alone make a basic knowledge of linear algebra relevant to machine learning. One example of using vectors is found in a process known as one hot encoding. There are a series of bits representing possible values for a specific piece of data. The only output that is a 1, is the value represented in the data, the other possible values are all 0. Consider for a moment that you need to import data that represents colors red, green and blue. You can use a vector that represents the color you are importing as a 1, and the other possible colors as 0. Thus, red would be 1,0,0 and green would be 0,1,0. You will find one hot encoding frequently used in machine learning.

Another topic that is common in machine learning, which depends on linear algebra, is principle component analysis (PCA). PCA is used in machine learning to create projections of high-dimensional data for both visualization and for training models. PCA, in turn, depends on matrix factorization. The Eigendecomposition is often used in PCA. Thus, to truly understand PCA, one needs to understand linear algebra.

Another application of linear algebra to machine learning combines linear algebra with statistics (which we will be reviewing in chapter 2). Linear regression is a statistical method for describing the relationship between variables. More specifically, linear regression describes the relationship between a variable and one or more explanatory variables.

Linear algebra was initially developed as a method to solve systems of linear equations (thus the name). However, we will move beyond solving linear equations.

DOI: 10.1201/9781003230588-2

Linear equations are those for which all elements are of the first power. Thus, the following three equations are linear equations:

$$a + b + c = 115$$

$$3x + 2y = 63$$

$$2x + y - z = 31$$

However, the following are not linear equations:

$$2x^2 + 3 = 12$$

$$4y^2 + 6x + 2 = 22$$

$$7x^4 + 8y^3 - 1 = 8$$

The first three equations have all individual elements only raised to the first power (often that with a number such as $x1$, the 1 is simply assumed and not written). But in the second set of equations, at least one element is raised to some power greater than 1. Thus, they are not linear. While linear algebra was created to solve linear equations, the subject has grown to encompass a number of mathematical endeavors that are not directly focused on linear equations. The reason for that is that linear algebra represents numbers in matrix form, which we shall examine in some detail in this chapter. That matrix form turns out to be ideal for a number of applications, including machine learning. Certainly, linear algebra is useful in machine learning, thus this chapter.

It may help to begin with a very brief history of linear algebra. One of the earliest books on the topic of linear algebra was *Extension Theory* written in 1844, by Hermann Grassman. The book included other topics, but also had some fundamental concepts of linear algebra. As time progressed, other mathematicians added additional features to linear algebra. In 1856, Arthur Cayle introduced matrix multiplication. Matrix multiplication is one of the elements of linear algebra that has applications beyond solving linear equations. In 1888, Giuseppe Peano gave a precise definition of a vector space. Vector spaces also have applications well beyond solving linear equations. Linear algebra continued to evolve over time.

This chapter is, by design, a brief overview of linear algebra. Engineering students typically take at least one entire undergraduate course in linear algebra. Mathematics majors may take additional course including graduate courses. Clearly, a single chapter in a book cannot cover all of that. But it is not necessary for you to learn linear algebra to that level in order to begin working with machine learning. The most obvious thing that is omitted in this chapter are proofs. We do not mathematically prove any of what is presented. While that might grate on the more mathematically oriented reader, the proofs are not necessary for our purposes. If you are interested in delving deeper into linear algebra, the Mathematical Association of America has a website with resources on linear algebra you can find at www.maa.org/topics/history-of-linear-algebra.

LINEAR ALGEBRA BASICS

As was discussed in the introduction, the goal of this chapter is not to solve linear equations. Rather, the goal is to provide the reader with sufficient understanding of linear algebra in order to apply that knowledge to machine learning. Machine learning frequently deals with matrices and vectors. Therefore, it is appropriate to begin this exposition of linear algebra with a discussion of matrices. A matrix is a rectangular arrangement of numbers in rows and columns. Rows run horizontally and columns run vertically. The dimensions of a matrix are stated m x n where m is the number of rows and n is the number of columns. Here is an example:

$$\begin{bmatrix} 1 & 2 \\ 2 & 0 \\ 3 & 1 \end{bmatrix}$$

If this definition seems a bit elementary to you, you are correct. This is a rather straightforward way to represent numbers. The basic concept of a matrix is not at all difficult to grasp. That is yet another reason it is an appropriate place to begin exploring linear algebra. A matrix is just an array that is arranged in columns and rows. Vectors are simply matrices that have one column or one row. The examples in this section focus on 2 x 2 matrices, but a matrix can be of any number of rows and columns, it need not be a square. A vector can be considered a 1 x m matrix. A vector that is vertical is called a column vector, one that is horizontal is called a row vector. Matrices are usually labeled based on column and row:

$$\begin{bmatrix} a_{ij} & a_{ij} \\ a_{ij} & a_{ij} \end{bmatrix}$$

The letter i represents the row, and the letter j represents the column. A more concrete example is shown here:

$$\begin{bmatrix} a_{11} & a_{12} \\ a_{21} & a_{22} \end{bmatrix}$$

This notation is commonly used for matrices including row and column vectors. In addition to understanding matrix notation, there are some common types of matrices you should be familiar with. The most common are listed here:

Column Matrix: A matrix with only one column.
Row Matrix: A matrix with only one row.
Square Matrix: A matrix that has the same number of rows and columns
Equal Matrices: Two matrices are considered equal if they have the same number of rows and columns (the same dimensions) and all their corresponding elements are exactly the same.
Zero matrix: Contains all zeros.

Each of these has a role in linear algebra, which you will see as you proceed through the chapter. Later in this book you will see column matrices used frequently as features being analyzed by machine learning algorithms.

MATRIX ADDITION AND MULTIPLICATION

Addition and multiplication of matrices are actually not terribly complicated. There are some basic rules you need to be aware of to determine if two matrices can be added or multiplied. If two matrices are of the same size, then they can be added to each other by simply adding each element together. You start with the first row and first column in the first matrix and add that to the first row and the first column of the second matrix thus in the sum matrix. This is shown in equation 1.1:

$$\begin{bmatrix} a_{11} & a_{12} \\ a_{21} & a_{22} \end{bmatrix} + \begin{bmatrix} b_{11} & b_{12} \\ b_{21} & b_{22} \end{bmatrix} = \begin{bmatrix} A_{11} + b_{11} & a_{12} + b_{12} \\ A_{21+}b_{21} & a_{22} + b_{22} \end{bmatrix} \qquad \text{(eq. 1.1)}$$

Consider the following more concrete example given in equation 1.2:

$$\begin{bmatrix} 1 & 2 \\ 2 & 2 \end{bmatrix} + \begin{bmatrix} 1 & 2 \\ 2 & 1 \end{bmatrix} = \begin{bmatrix} 2 & 4 \\ 4 & 3 \end{bmatrix} \qquad \text{(eq. 1.2)}$$

This is rather trivial to understand, thus this journey into linear algebra begins with an easily digestible concept. Multiplication, however, is somewhat more difficult. One can only can multiply two matrices if the number of columns in the first matrix is equal to the number of rows in the second matrix. First let us take a look at multiplying a matrix by a scalar (i.e., a single number). You simply multiply the scalar value by each element in the matrix as shown in equation 1.3.

$$C\begin{bmatrix} a_{ij} & a_{ij} \\ a_{ij} & a_{ij} \end{bmatrix} = \begin{bmatrix} ca_{ij} & ca_{ij} \\ ca_{ij} & ca_{ij} \end{bmatrix} \qquad \text{(eq. 1.3)}$$

For a more concrete example, consider equation 1.4:

$$3\begin{bmatrix} 1 & 1 \\ 2 & 3 \end{bmatrix} = \begin{bmatrix} 3 & 3 \\ 6 & 9 \end{bmatrix} \qquad \text{(eq. 1.4)}$$

As you can observe for yourself, the addition of two matrices is no more complicated than the addition you learned in primary school. However, the multiplication of two matrices is a bit more complex. The two matrices need not be of the same size. The requirement is that the number of columns in the first matrix be equal to the number of rows in the second matrix. If that is the case, then you multiply each element in the first row of the first matrix, by each element in the second matrix's first column. Then you multiply each element of the second row of the first matrix by each element of the second matrix's second column. Let's first examine this using variables rather than actual numbers. This example also uses square matrices to make the situation even simpler. This is shown in equation 1.5.

$$\begin{bmatrix} a & b \\ c & d \end{bmatrix} + \begin{bmatrix} e & f \\ g & h \end{bmatrix}$$

(eq. 1.5)

This is multiplied in the following manner.

$$\begin{aligned} a*e+b*g & \quad (a_{11}*b_{11}+a_{12}*b_{21}) \\ a*f+b*h & \quad (a_{11}*b_{12}+a_{12}*b_{22}) \\ c*e+d*g & \quad (a_{11}*b_{11}+a_{12}*b_{21}) \\ c*f+d*h & \quad (a_{11}*b_{11}+a_{12}*b_{21}) \end{aligned}$$

Thus, the product will be:

$$\begin{aligned} (a*e+b*g) & \quad (a*f+b*h) \\ (c*e+d*g) & \quad (c*f+d*h) \end{aligned}$$

It is worthwhile to memorize this process. Now, consider this implemented with a concrete example, as shown in equation 1.6:

$$\begin{bmatrix} 1 & 3 \\ 3 & 1 \end{bmatrix} \begin{bmatrix} 2 & 2 \\ 1 & 3 \end{bmatrix}$$

(eq. 1.6)

We begin with:

$$\begin{aligned} 1*2+3*1 &= 5 \\ 1*2+3*3 &= 11 \\ 3*2+1*1 &= 7 \\ 3*2+1*3 &= 9 \end{aligned}$$

The final answer is:

$$\begin{bmatrix} 5 & 11 \\ 7 & 9 \end{bmatrix}$$

It should first be noted that one can only multiply two matrices if the number of columns in the first matrix equals the number of rows in the second matrix. That should be clear if you reflect on how the multiplication is done. If you keep that rule in mind, then the multiplication is not particularly complicated, it is simply tedious.

It is important to remember that matrix multiplication, unlike multiplication of scalers, is not commutative. For those readers who may not recall, the commutative property states $a * b = b * a$. If a and b are scalar values then this is true, regardless of the scaler values (e.g., integers, rational numbers, real numbers, etc.). However, when multiplying two matrices, this is not the case. This is often a bit difficult for those new to matrix mathematics. However, it is quite easy to demonstrate that this commutative property does not hold for matrices. For example, consider the matrix multiplication shown in Equation 1.7.

$$\begin{bmatrix} 2 & 1 \\ 3 & 2 \end{bmatrix}\begin{bmatrix} 3 & 2 \\ 2 & 1 \end{bmatrix} = \begin{bmatrix} 8 & 5 \\ 13 & 8 \end{bmatrix}$$

(eq. 1.7)

However, if you reverse that order you will get a different answer. This is shown in equation 1.8.

$$\begin{bmatrix} 3 & 2 \\ 2 & 1 \end{bmatrix}\begin{bmatrix} 2 & 1 \\ 3 & 2 \end{bmatrix} = \begin{bmatrix} 12 & 7 \\ 7 & 4 \end{bmatrix}$$

(eq. 1.8)

Seeing an actual example, it becomes immediately apparent that matrix multiplication is not commutative. You may find some instance that just incidentally appears communitive. However, that is not sufficient. To have the commutative property, it must be the case that, regardless of the operands chosen, the result is commutative.

OTHER MATRIX OPERATIONS

Matrix addition and multiplication are perhaps the most obvious of operations on matrices. These operations are familiar to you from previous mathematics you have studied. However, there are some operations that are specific to matrices. Matrix transposition is one such operation. Transposition simply reverses the order of rows and columns. While the focus so far has been on 2 x 2 matrices, the transposition operation is most easily seen with a matrix that has a different number of rows and columns. Consider the matrix shown in Equation 1.9.

$$\begin{bmatrix} 2 & 3 & 2 \\ 1 & 4 & 3 \end{bmatrix}$$

(eq. 1.9)

To transpose it, the rows and columns are switched, creating a 2 x 3 matrix. The first row is now the first column. You can see this in Equation 1.10.

$$\begin{bmatrix} 2 & 1 \\ 3 & 4 \\ 2 & 3 \end{bmatrix}$$

(eq. 1.10)

There are some particular properties of transpositions that you should be familiar with. If you label the first matrix A, then the transposition of that matrix is labeled AT. Continuing with the original matrix being labeled A, there are a few properties of matrices that need to be described as outlined in Table 1.1.

TABLE 1.1

Basic Properties of Matrix Transposition

Property	Explanation
$(A^T)^T = A$	If you transpose the transposition of A, you get back to A.
$(cA)^T = cA^T$	The transposition of a constant, c, multiplied by an array, A, is equal to multiplying the constant c by the transposition of A.
$(AB)^T = B^T A^T$	A multiplied by B, then the product transposed, is equal to B transposed multiplied by A transposed.
$(A + B)^T = A^T + B^T$	Adding the matrix A and the matrix B, then transposing the sum, is equal to first transposing A and B, then adding those transpositions.
$A^T = A$	If a square matrix is equal to its transpose, it is called a symmetric matrix.

Table 1.1 is not exhaustive, rather, it is a list of some of the most common properties regarding matrices. These properties are not generally particularly difficult to understand. Another operation particular to matrices is finding a submatrix of a given matrix. A submatrix is any portion of a matrix that remains after deleting any number of rows or columns. Consider the 5 x 5 matrix shown in Equation 1.11

$$\begin{bmatrix} 2 & 2 & 4 & 5 & 3 \\ 3 & 8 & 0 & 2 & 1 \\ 2 & 3 & 2 & 2 & 1 \\ 4 & 3 & 1 & 2 & 4 \\ 1 & 2 & 2 & 0 & 3 \end{bmatrix} \quad \text{(eq. 1.11)}$$

If you remove the second column and second row, as shown in Equation 1.12:

$$\begin{bmatrix} 2 & 2 & 4 & 5 & 3 \\ 3 & 8 & 0 & 2 & 1 \\ 2 & 3 & 2 & 2 & 1 \\ 4 & 3 & 1 & 2 & 4 \\ 1 & 2 & 2 & 0 & 3 \end{bmatrix} \quad \text{(eq. 1.12)}$$

You are left with the matrix shown in Equation 1.13.

$$\begin{bmatrix} 2 & 4 & 5 & 3 \\ 2 & 2 & 2 & 1 \\ 4 & 1 & 2 & 4 \\ 1 & 2 & 0 & 3 \end{bmatrix} \quad \text{(eq. 1.13)}$$

That matrix shown in Equation 1.13 is a sub matrix of the original matrix.

Another item that is important to matrices is the identity matrix. Now this item does have an analog in integers and real numbers. Regarding the addition operation, the identity element is 0. Any number + 0 will still be the original number. In multiplication, the identity element is 1. Any number * 1 is still the original number. The identity matrix functions in much the same manner. Multiplying a matrix by its identity matrix leaves it unchanged. To create an identity matrix, just have all the elements along the main diagonal set to 1, and the rest to zero. Consider the following matrix:

$$\begin{bmatrix} 3 & 2 & 1 \\ 1 & 1 & 2 \\ 3 & 0 & 3 \end{bmatrix}$$

Now consider the identity matrix. It must have the same number of columns and rows, with its main diagonal set to all 1s and the rest of the elements all 0s. The identity matrix looks like this:

$$\begin{bmatrix} 1 & 0 & 0 \\ 0 & 1 & 0 \\ 0 & 0 & 1 \end{bmatrix}$$

If you multiply the original matrix by the identity matrix, the product will be the original matrix. You can see this in Equation 1.14

$$\begin{bmatrix} 3 & 2 & 1 \\ 1 & 1 & 2 \\ 3 & 0 & 3 \end{bmatrix} \times \begin{bmatrix} 1 & 0 & 0 \\ 0 & 1 & 0 \\ 0 & 0 & 1 \end{bmatrix} = \begin{bmatrix} 3 & 2 & 1 \\ 1 & 1 & 2 \\ 3 & 0 & 3 \end{bmatrix} \qquad \text{(eq. 1.14)}$$

Another special type of matrix is a unimodular matrix. Unimodular matrices are also used in some lattice-based algorithms. A unimodular matrix is a square matrix of integers with a determinant of +1 or -1. Recall that a determinant is a value that is computed from the elements of a square matrix. The determinant of a matrix A is denoted by |A|. In the next section we will explore how to calculate the determinant of a matrix.

DETERMINANT OF A MATRIX

Next, we will turn our attention to another relatively easy computation, the determinant of a matrix. The determinant of a matrix A is denoted by |A|. An example of a determinant in a generic form is as follows:

$$|A| \begin{bmatrix} a & b \\ C & d \end{bmatrix} = ad - bc$$

A more concrete example might help elucidate this concept:

$$|A| \begin{bmatrix} 2 & 3 \\ 1 & 2 \end{bmatrix} = (2)(2) - (3)(1) = 1$$

A determinant is a value that is computed from the individual elements of a square matrix. It provides a single number, also known as a scaler value. Only a square matrix can have a determinant. The calculation for a 2 x 2 matrix is simple enough, we will explore more complex matrices in just a moment. However, what does this single scalar value mean? There are many things one can do with a determinant, most of which we won't use in this text. It can be useful in solving linear equations, changing variables in integrals (yes, linear algebra and calculus go hand in hand); however, what is immediately useable for us is that if the determinant is nonzero, then the matrix is invertible.

What about a 3 x 3 matrix, such as that shown in Equation 1.15?

$$\begin{bmatrix} a_1 & b_1 & c_1 \\ a_2 & b_2 & c_2 \\ a_3 & b_3 & c_3 \end{bmatrix} \qquad \text{(eq. 1.15)}$$

This calculation is substantially more complex. There are a few methods to do this. We will use one called 'expansion by minors'. This method depends on breaking the 3 x 3 matrix into 2 x 2 matrices. The 2 x 2 matrix formed by b_2, c_2, b_3, c_3, shown in Equation 1.16, is the first.

$$\begin{bmatrix} a_1 & b_1 & c_1 \\ a_2 & b_2 & c_2 \\ a_3 & b_3 & c_3 \end{bmatrix} \qquad \text{(eq. 1.16)}$$

This one was rather simple, as it fits neatly into a contiguous 2 x 2 matrix. But to find the next one, we have a bit of a different selection as shown in Equation 1.17:

$$\begin{bmatrix} a_1 & b_1 & c_1 \\ a_2 & b_2 & c_2 \\ a_3 & b_3 & c_3 \end{bmatrix} \qquad \text{(eq. 1.17)}$$

The next step is to get the lower left corner square matrix as shown in Equation 1.18.

$$\begin{bmatrix} a_1 & b_1 & c_1 \\ a_2 & b_2 & c_2 \\ a_3 & b_3 & c_3 \end{bmatrix} \qquad \text{(eq. 1.18)}$$

As with the first one, this one forms a very nice 2 x 2 matrix. Now what shall we do with these 2 x 2 matrices? The formula is actually quite simple and is shown in Equation 1.19. Note that det is simply shorthand for determinant.

$$\det \begin{bmatrix} a_1 & b_1 & c_1 \\ a_2 & b_2 & c_2 \\ a_3 & b_3 & c_3 \end{bmatrix} = a_1 \det \begin{bmatrix} b_2 & c_2 \\ b_3 & c_3 \end{bmatrix} - a_2 \det \begin{bmatrix} a_2 & c_2 \\ a_3 & c_3 \end{bmatrix} + a_3 \det \begin{bmatrix} a_2 & b_2 \\ a_3 & b_3 \end{bmatrix} \quad \text{(eq. 1.19)}$$

We take the first column, multiplying it by its cofactors, and with a bit of simple addition and subtraction, we arrive at the determinant for a 3 x 3 matrix. A more concrete example might be useful. Let us calculate the determinant for this matrix:

$$\begin{bmatrix} 3 & 2 & 1 \\ 1 & 1 & 2 \\ 3 & 0 & 3 \end{bmatrix}$$

This leads to:

$$3 * \det \begin{bmatrix} 1 & 2 \\ 0 & 3 \end{bmatrix} = 3 * ((1*3) - (2*0)) = 3(3) = 9$$

$$2 * \det \begin{bmatrix} 1 & 2 \\ 3 & 3 \end{bmatrix} = 2 * ((1*3) - (2*3)) = 2(-3) = -6$$

$$1 * \det \begin{bmatrix} 1 & 1 \\ 3 & 0 \end{bmatrix} = 1 * (((1*0) - (1*3)) = 1(-3) = -3$$

And that leads us to $9 - (-6) + (-3) = 12$

Yes, that might seem a bit cumbersome, but the calculations are not overly difficult. We will end our exploration of determinants at 3 x 3 matrices. But yes, one can take the determinant of larger square matrices. One can calculate determinants for matrices that are 4 x 4, 5 x 5 and as large as you like. However, our goal in this chapter is to give you a bit of general foundation in linear algebra, not to explore every nuance of linear algebra.

VECTORS AND VECTOR SPACES

Vectors are an essential part of linear algebra. We normally represent data in the form of vectors. In linear algebra, these vectors are treated like numbers. They can be added and multiplied. A vector will look like what is shown here:

$$\begin{bmatrix} 1 \\ 3 \\ 2 \end{bmatrix}$$

This vector has only integers; however, a vector can have rational numbers, real numbers, even complex numbers (which will be discussed later in this chapter). Vectors can also be horizontal, as shown here:

$$\begin{bmatrix} 1 & 3 & 2 \end{bmatrix}$$

Often you will see variables in place of vector numbers, such as:

$$\begin{bmatrix} a \\ b \\ c \end{bmatrix}$$

The main point that that you saw in the previous section is that one can do math with these vectors as if they were numbers. You can multiple two vectors together; you can also multiply a vector by a scaler. Scalers are individual numbers, and their name derives from the fact that they change the scale of the vector. Consider the scaler 3 multiplied by the first vector shown in this section:

$$3\begin{bmatrix} 1 \\ 3 \\ 2 \end{bmatrix} = \begin{bmatrix} 3 \\ 9 \\ 6 \end{bmatrix}$$

You simply multiply the scaler, by each of the elements in the vector. We will be exploring this and other mathematical permutations in more detail in the next section. But let us address the issue of why it is called a scaler now. We are viewing the data as a vector; another way to view it would be as a graph. Consider the previous vector [1,3,2] on a graph as shown in Figure 1.1.

Now what happens when we perform the scalar operation of 3 multiplied by that vector? We literally change the scale of the vector, as shown in Figure 1.2.

Figures 1.1 and 1.2 may appear identical, but look closer. In figure 1.1, the *x* value goes to 9, whereas in figure 1.2 the *x* value only goes to 3. We have 'scaled' the vector. The term scalar is used because it literally changes the scale of the vector. Formally, a vector space is a set of vectors that is closed under addition and multiplication by real numbers. Think back to the earlier discussion of abstract algebra with groups, rings and fields. A vector space is a group. In fact, it is an abelian group. You can do addition of vectors, and the inverse. You also have a second operation scaler multiplication, without the inverse. Note that the first operation, addition, is commutative, but the second operation, multiplication, is not.

So, what are basis vectors? If you have a set of elements *E* (i.e., vectors) in some vector space *V*, the set of vectors *E* is considered a basis if every vector in the vector space *V* can be written as a linear combination of the elements of *E*. Put another way, you could begin with the set *E*, the basis, and through a linear combinations of the vectors in *E*, create all the vectors in the vector space *V*. And as the astute reader will have surmised, a vector space can have more than one basis set of vectors.

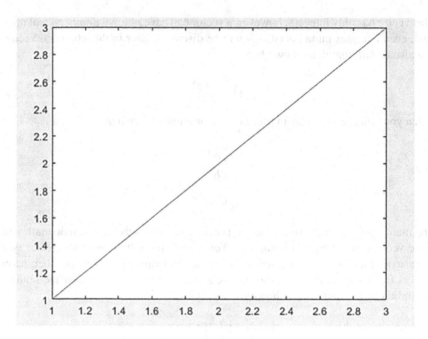

FIGURE 1.1   Graph of a vector.

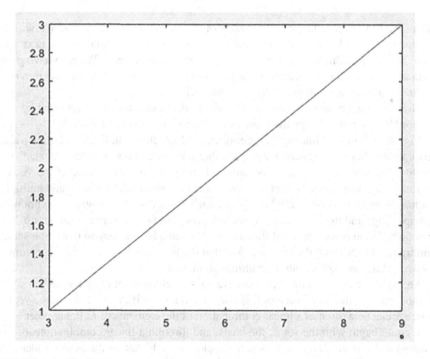

FIGURE 1.2   Scaling a vector.

What is linear dependence and independence? In the theory of vector spaces, a set of vectors is said to be linearly dependent if one of the vectors in the set can be defined as a linear combination of the others; if no vector in the set can be written in this way, then the vectors are said to be linearly independent.

A subspace is a subset of a vector space that is a vector space itself, e.g., the plane $z = 0$ is a subspace of R3 (it is essentially R2.). We'll be looking at Rn and subspaces of Rn.

VECTOR METRICS

There are a number of metrics that one can calculate from vector. Each of these is important in some application of linear algebra. In this section we will cover the most common metrics that you will encounter when applying matrix mathematics.

VECTOR LENGTH

Let us begin this section with a fairly easy topic, the length of a vector, which is computed using the Pythagorean theorem:

$$\|\text{vector}\| = \sqrt{x2} + y2$$

Consider the vector:

$$[2,3,4]$$

Its length is:

$$\sqrt{2^2 + 3^2 + 4^2} = 5.38$$

This is simple but will be quite important as we move forward. Now we will add just a bit more detail to this concept. The nonnegative length is called the norm of the vector. Given a vector v, this is written as ‖v‖. This will be important later on in this book. One more concept to remember on lengths/norms: if the length is 1, then this is called a unit vector.

DOT PRODUCT

The dot product of two vectors has numerous applications. This is an operation you will likely encounter with some frequency. The dot product of two vectors is simply the two vectors multiplied. Consider vectors X and Y. Equation 1.20 shows what the dot product would be.

$$\sum_{i=1}^{n} X_i Y_i \qquad \text{(eq. 1.20)}$$

Examining a concrete example to see how this works should be helpful. Consider two column vectors:

$$\begin{bmatrix} 1 \\ 2 \\ 1 \end{bmatrix} \begin{bmatrix} 3 \\ 2 \\ 1 \end{bmatrix}$$

The dot product is found by $(1 * 3) + (2 * 2) + (1 * 1) = 8$

That is certainly an easy calculation to perform. But what does it mean? Put more frankly, why should you care what the dot product is? Recall that vectors can also be described graphically. You can use the dot product, along with the length of the vectors to find the angle between the two vectors. We already know the dot product is 8. Recall that length:

$$\|\text{vector}\| = \sqrt{x2} + y2$$

Thus, the length of vector X is $\sqrt{1^2 + 2^2 + 1^2} = 2.45$

The length of vector Y is $\sqrt{3^2 + 2^2 + 1^2} = 3.74$

Now we can easily calculate the angle. It turns out that the $\cos \theta$ = dot product/ length of X * length of Y

or

$$\cos \theta = \frac{8}{(2.45)(3.74)} = .87307$$

Finding the angle from the cosine is straightforward; you probably did this in secondary school trigonometry. But even with just the dot product, we have some information. If the dot product is 0 then the vectors are perpendicular. This is because the $\cos \theta$ of a 90-degree angle is 0. The two vectors are referred to as orthogonal.

Recall that the length of a vector is also called the vector's norm. And if that length/norm is 1, it is the unit vector. This leads us to another term we will see frequently later in this book. If two vectors are both orthogonal (i.e., perpendicular to each other) and have unit length (length 1), the vectors are said to be orthonormal.

Essentially, the dot product is used to produce a single number, a scaler, from two vertices or two matrices. This is contrasted with the tensor product. In math, a tensor is an object with multiple indices, such as a vertex or array. The tensor product of two vector spaces, V and W, $V \otimes W$ is also a vector space.

TENSOR PRODUCT

This is essentially a process where all of the elements from the first vector are multiplied by all of the elements in the second vector. This is shown in figure 1.3:

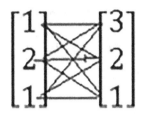

FIGURE 1.3   Tensor Product.

CROSS PRODUCT

This particular metric is very interesting. It illustrates the fact that vector mathematics is inherently geometric. Even if you are working only with vectors, and not seeing the graphical representation, it is there. Let us first describe how you calculate the cross product, then discuss the geometric implications.

If you have two vectors A and B, the cross product is defined as:

$$A \times B = |A| \, \|B| \sin(\theta) n$$

$|A|$ denotes the length of A (note: length is often called magnitude)

$|B|$ denotes the length of B

N is the unit vector that is at right angles to both A and B

θ is the angle between A and B

If you reflect on this a bit, you will probably notice that this requires a three-dimensional space. Using that fact, there is a simple way to calculate the cross product C:

$$cx = A_y B_z - A_z B_y$$

$$cy = A_z B_x - A_x B_z$$

$$cz = A_x B_x - A_y B_x$$

So, consider two vectors: $A = [2, 3, 4]$ and $B = [5, 6, 7]$.

$$Cx = 3 * 7 - 4 * 6 = 21 - 24 = -3$$

$$Cy = 4*5 - 2*7 = 20 - 14 = 6$$

$$Cz = 2*6 - 3*5 = 12 - 15 = -3$$

The cross product is [-3, 6, -3].

EIGENVALUES AND EIGENVECTORS

Eigenvalues are a special set of scalars associated with a linear system of equations (i.e., a matrix equation) that are sometimes also known as characteristic roots, characteristic values, proper values, or latent roots. To clarify, consider a column vector we will call v. Then also consider an n x n matrix we will call A. Then consider some scalar λ. If it is true that:

$$Av = \lambda v$$

Then we say that v is an eigenvector of the matrix A and λ is an eigenvalue of the matrix A.

Let us look a bit closer at this. The prefix eigen is actually the German word which can be translated as specific, proper, particular, etc. Put in its most basic form, an eigenvector of some linear transformation T, is a vector that, when T is applied to it, does not change direction, it only changes scale. It changes scale by the scalar value λ, the eigenvalue. Now we can revisit the former equation just a bit and expand our knowledge of linear algebra:

$$T(v) = \lambda v$$

This appears precisely like the former equation, but with one small difference. The matrix A is now replaced with the transformation T. Not only does this tell us about eigenvectors and eigenvalues, but it also tells us a bit more about matrices. A matrix, when applied to a vector, transforms that vector. The matrix itself is an operation on the vector. This is a concept that is fundamental to matrix theory.

Let us add something to this. How do you find the eigenvalues and eigenvectors for a given matrix? Surely it is not just a matter of trial and error with random numbers. Fortunately, there is a very straightforward method, one that is actually quite easy, at least for 2 x 2 matrices. Consider the following matrix:

$$\begin{bmatrix} 5 & 2 \\ 9 & 2 \end{bmatrix}$$

How Do We Find Its Eigenvalues?

Well, the Cayley-Hamilton theorem provides insight on this issue. That theorem essentially states that a linear operator A is a zero of its characteristic polynomial. For our purposes, it means that

$$\det |A - \lambda I| = 0$$

We know what a determinant is, we also know that I is the identity matrix. The λ is the eigenvalue we are trying to find. The A is the matrix we are examining. Remember that in linear algebra you can apply a matrix to another matrix or vector, so a matrix is, at least potentially, an operator. So, we can fill in this equation:

$$\det\left\|\begin{bmatrix} 5 & 2 \\ 9 & 2 \end{bmatrix} - \lambda\begin{bmatrix} 1 & 0 \\ 0 & 1 \end{bmatrix}\right\| = 0$$

Now, we just have to do a bit of algebra, beginning with multiplying λ by our identity matrix, which will give us:

$$\det\left|\begin{bmatrix} 5 & 2 \\ 9 & 9 \end{bmatrix} - \begin{bmatrix} \lambda & 0 \\ 0 & \lambda \end{bmatrix}\right| = 0$$

Which in turn leads to

$$\det\left\|\begin{bmatrix} 5-\lambda & 2 \\ 9 & 2-\lambda \end{bmatrix}\right\| = 0$$
$$= (5-\lambda)(5-\lambda) - 18$$
$$= 10 - 7\lambda + \lambda^2 - 18 = 0$$
$$\lambda^2 - 7\lambda \quad -8 = 0$$

This can be factored (note if the result here cannot be factored, things do get a bit more difficult, but that is beyond our scope here):

$$(\lambda - 8)(\lambda + 1) = 0$$

This means we have two eigenvalues:

$$\lambda_1 = 8$$

$$\lambda_2 = -1$$

For a 2 x 2 matrix you will always get two eigenvalues. In fact, for any n x n matrix, you will get n eigenvalues, but they may not be unique.

Now that you have the eigenvalues, how do you calculate the eigenvectors?
We know that:

$$A = \begin{bmatrix} 5 & 2 \\ 9 & 2 \end{bmatrix}$$

$$\lambda_1 = 8$$

$$\lambda_2 = -1$$

We are seeking unknown vectors, so let us label the vector $\begin{bmatrix} X \\ Y \end{bmatrix}$.

Now recall the equation that gives us eigenvectors and eigenvalues:

$$Av = \lambda v$$

Let us take one of our eigenvalues and plug it in:

$$\begin{bmatrix} 5 & 2 \\ 9 & 2 \end{bmatrix}\begin{bmatrix} X \\ Y \end{bmatrix} = 8\begin{bmatrix} X \\ Y \end{bmatrix}$$

$$\begin{bmatrix} 5x+2y \\ 9x+2y \end{bmatrix} = \begin{bmatrix} 8X \\ 8Y \end{bmatrix}$$

This gives us two equations:

$$5x + 2y = 8x$$

$$9x + 2y = 8y$$

Now we take the first equation and do a bit of algebra to isolate the y value. Subtract the $5x$ from each side to get:

$$2y = 3x$$

Then divide both sides by 2 to get

$$y = 3 / 2x$$

It should be easy to see that to solve this with integers (which is what we want), then $x = 2$ and $y = 3$ solve it. Thus, our first eigenvector is

$$\begin{bmatrix} 2 \\ 3 \end{bmatrix} \text{ with } \lambda_1 = 8$$

You can work out the other eigenvector for the second eigenvalue on your own using this method.

EIGENDECOMPOSITION

As was mentioned at the beginning of this chapter, Eigendecomposition is key to principle component analysis. The process of Eigendecomposition is the factorization of a matrix into what is called canonical form. That means that the matrix is represented in terms of its eigenvalues and eigenvectors.

Consider a square matrix A that is n x n with n linearly independent eigenvectors q_i (where $i = 1, \ldots n$). If this is true, then the matrix A can be factorized as shown in equation 1.21.

$$A = Q \wedge A - 1 \qquad \text{(eq. 1.21)}$$

In equation 1.21, Q is a square n x n matrix whose ith column is the eigenvector q_i of A. The symbol $\wedge$ is a diagonal matrix whose diagonal elements are the corresponding eigenvalues. It should be noted that if a matrix A can be eigendecomposed and if none of its eigenvalues are zero, then that matrix is invertible. For more information on Eigendecomposition, the following sources may be of use:

https://mathworld.wolfram.com/EigenDecomposition.html
https://personal.utdallas.edu/~herve/Abdi-EVD2007-pretty.pdf

SUMMARY

Linear algebra is an important topic for machine learning. The goal of this chapter is to provide you with a basic introduction to linear algebra, or perhaps for some readers a review. You should ensure that you have mastered these concepts before proceeding to subsequent chapters. That means you should be comfortable adding and multiplying matrices, calculating dot products, calculating the determinant of a matrix, and working with eigenvalues and eigenvectors. You will see eigenvalues and eigenvectors used again in relation to spectral graph theory in chapter 8. These are fundamental operations in linear algebra. The multiple-choice questions will also help to ensure you have mastered the material. If you wish to go further with linear algebra, you may find the following resources useful:

McMahon, D. 2005. *Linear algebra demystified*. McGraw Hill Professional.
    This is a good resource for basic linear algebra.
Aggarwal, Charu. 2020. *Linear algebra and optimization for machine learning* (Vol. 156). Springer International Publishing.
Schneider, H. and Barker, G. P. 2012. *Matrices and linear algebra (Dover books on mathematics)*. Dover Publications.

TEST YOUR KNOWLEDGE

1. Solve the equation $2 \begin{bmatrix} 2 \\ 3 \\ 4 \end{bmatrix}$.

    a. $\begin{bmatrix} 4 \\ 6 \\ 8 \end{bmatrix}$

b. $\begin{bmatrix} 5 \\ 4 \\ 6 \end{bmatrix}$

c. 18

d. 15

2. What is the dot product of these two vectors?

$$\begin{bmatrix} 1 \\ 2 \\ 3 \end{bmatrix} \begin{bmatrix} 4 \\ 5 \\ 6 \end{bmatrix}$$

a. 11

b. 32

c. 28

d. 21

3. Solve this determinant $|A| \begin{bmatrix} 2 & 2 \\ 3 & 4 \end{bmatrix}$.

a. 8

b. 6

c. 0

d. –2

4. Solve this determinant: $|A| \begin{bmatrix} 1 & 2 & 3 \\ 2 & 1 & 4 \\ 3 & 1 & 2 \end{bmatrix}$.

a. 11

b. 12

c. 15

d. 10

5. What is the dot product of these two vectors?

$$\begin{bmatrix} 1 \\ 2 \\ 2 \end{bmatrix} \begin{bmatrix} 3 \\ 3 \\ 4 \end{bmatrix}$$

a. 65

b. 17

c. 40

d. 15

6. What is the length of this vector? $\begin{bmatrix} 3 \\ 3 \\ 2 \end{bmatrix}$

    a. 4.35
    b. 2.64
    c. 3.46
    d. 4.56

7. What is the product of these two matrices? $\begin{bmatrix} 2 & 3 \\ 1 & 4 \end{bmatrix} \begin{bmatrix} 3 & 2 \\ 3 & 7 \end{bmatrix}$

    a. $\begin{bmatrix} 15 & 25 \\ 15 & 10 \end{bmatrix}$

    b. $\begin{bmatrix} 15 & 10 \\ 15 & 30 \end{bmatrix}$

    c. $\begin{bmatrix} 10 & 15 \\ 15 & 30 \end{bmatrix}$

    d. $\begin{bmatrix} 15 & 25 \\ 15 & 30 \end{bmatrix}$

8. Are these two vectors orthogonal?

$$\begin{bmatrix} 1 \\ 0 \end{bmatrix} \begin{bmatrix} 0 \\ 1 \end{bmatrix}$$

    a. Yes
    b. No

9. Write the identity matrix for this matrix:

$$\begin{bmatrix} 3 & 2 & 1 \\ 3 & 5 & 6 \\ 3 & 4 & 3 \end{bmatrix}$$

Answer:

$$\begin{bmatrix} 1 & 0 & 0 \\ 0 & 1 & 0 \\ 0 & 0 & 1 \end{bmatrix}$$

2 Overview of Statistics

INTRODUCTION

Machine learning often involves the use of statistics. And being completely candid, the more you know about statistics the better you will be able to utilize machine learning. This is not to imply that machine learning is merely the application of statistics. Unfortunately, some vendors who market alleged machine learning are merely doing statistics, but that is not true machine learning. In this chapter, the goal will be to familiarize yourself with the essential statistical knowledge you will need for effectively utilizing machine learning. For some readers this may be a review. For others, this may be your first foray into statistics.

Most texts on statistics begin with defining some terms, or jumping into basic formulas. And those shall certainly be covered in this chapter. However, we will begin with a more fundamental issue. Let us begin by exploring what statistics actually is. Statistics is a branch of mathematics designed to allow one to accomplish two goals. The first is to accurately describe data and trends in data. The second is to make predictions on future behavior, based on current data. The first goal is referred to as descriptive statistics. Any method or formula which yields some number that tells you about a set of data is considered descriptive statistics. Any method or formula which discusses a probability of some event occurring is predictive statistics.

In this chapter we will begin by discussing descriptive statistics. The goal is to summarize the current level of understanding of basic descriptive statistics and to give some general guidelines for using descriptive statistics. Later in the chapter, the basics of probability will be covered.

BASIC TERMINOLOGY

Before we can proceed, certain terminology must be covered. Without a thorough understanding of these terms, it is impossible for any person to be able to study even rudimentary statistics.

Continuous Variable: A quantitative variable that can assume an uncountable number of values. This means that a continuous variable can assume any value along a line interval, including every possible value between any two values.

Data (singular): The value of the variable associated with one element of a population or sample. This value may be a number, a word, or a symbol.

Data (plural): The set of values collected for the variable from each of the elements belonging to the sample.

Descriptive statistics: This is a common application of statistics. This process involves the classification, analysis, and interpretation of data.

DOI: 10.1201/9781003230588-3

Discrete Variable: A quantitative variable that can assume a countable number of values. This means that a discrete variable can only assume values corresponding to isolated points along a line interval.

Experiment: A planned activity whose results yield a set of data.

Hypothesis: The idea you are testing. Statistics are usually done in an attempt to confirm or refute some idea. Often in statistics you confirm or refute the null hypothesis, denoted as H_0. It is the hypothesis that essentially the results you get are random and are not due to some real relationship. In other words, if the null hypothesis is true, then the apparent relationship is really simply a random coincidence.

Predictive statistics: Using statistics generated from the sample in order to make predictions; this is also sometimes referred to as *inferential statistics*. This use of statistics deals with probabilities.

Parameter: This is a descriptive number about a population. A statistic is a descriptive number about a sample.

Population: The target group you wish to study, such as all men aged 30 to 40.

Sample: The subgroup from the population you select to study, in order to make inferences about the population.

Variable: A characteristic about each individual element of a population or sample.

These are basic terms, and it seems quite likely that many readers are already quite familiar with them. However, it is necessary to establish this basic lexicon before proceeding. Other terms will be defined as needed.

TYPES OF MEASUREMENT SCALES

When measuring, the scale you use determines how accurate you can be. In some instances, it is simply not possible to get beyond a certain level of accuracy. It is important to understand the scale you are using. You will find that, often in machine learning, one can deal with a wide range of data types.

1. Nominal: For qualitative data with distinct categories. For example, the categories of neurological imaging, such as MRI, PET, CAT, are categories but are not ordered in any way. Similarly, the analysis of such images to be tumor, injury, or healthy are categories, but without a specific order.
2. Ordinal: For qualitative data with distinct categories in which ordering (or ranking) is implied. A good example is the Likert scale that you see on many surveys: 1 = Strongly disagree; 2 = Disagree; 3 = Neutral; 4 = Agree; 5 = Strongly agree.
3. Interval: For quantitative data with an ordered scale in which the interval between data values is meaningful. For example, the categories of rank in the military. Clearly a major is higher-ranked than a captain, but how much higher? Does he have twice the authority of a captain? It is impossible to say. You can only say he is higher-ranked.

4. Ratio: For quantitative data that have an inherently defined zero and the ratio of data values is meaningful. Weight in kilograms is a very good example since it has a definite ratio from one weight to another. 50 kg is indeed twice as heavy as 25 kg. Clearly, when measuring the accuracy of a machine learning algorithm, you will want to have a ratio scale.

DATA COLLECTION

Normally, statistics are done with only a fraction of the actual population being considered. That fraction is called the sample, and the group in question is the population. For example, if you wish to examine the accuracy of a given neurological diagnostic process, you will require a sample of patients that have been through the process.

This leads to two obvious questions. Is the sample size you selected large enough and is the sample truly representative of the population you are attempting to measure? The first question will often be a controversial one. Obviously the larger the sample size the better. However, it is often impractical to get very large sample sizes. For example, in medical studies one cannot practically examine every patient that has a particular disease. Rather, one has to take a sample of patients that fit particular parameters. And, of course, with human subjects, the subject's willingness to participate is of paramount importance.

The second question, whether or not your sample is actually representative of the population you are trying to measure, is much easier to answer. There are some very specific ways in which you should select a sample. Using proper sampling techniques will give your statistical analysis credibility. The Statistics Glossary[1], lists several sampling techniques; each is described here:

Independent Sampling: This occurs when multiple samples are taken, but each sample has no effect on any other.

Random Sampling: This occurs when subjects for your sample are picked totally at random with no other factors influencing their selection. As an example, when names are drawn from a hat, one has random sampling.

Stratified Random Sampling: In this process, the population is divided into layers based on some criteria, and a number of random subjects are taken from each stratum. In our example of studying men who are over 40 and overweight, you might break the population into strata based on how overweight they are, or how old they are. For example, you might have men that are 25 to 50 lbs. overweight in one stratum and those who are 50 to 100 lbs. overweight in another, then finally those who are more than 100 lbs. over weight.

There are other sampling methods, but those just discussed are very commonly used. If you wish to learn more about sampling methods, the following websites will be helpful:

- Stat Pac, www.statpac.com/surveys/sampling.htm
- Statistics Finland, www.stat.fi/tk/tt/laatuatilastoissa/lm020500/pe_en.html

- Australian Bureau of Statistics, www.abs.gov.au/websitedbs/D3310116.
  NSF/4a255eef008309e44a255eef00061e57/116e0f93f17283eb4a2567ac00
  213517!OpenDocument

When evaluating any statistical analysis, it is important to consider how the sampling was done, and if the sample size seems large enough to be relevant. It might even be prudent to never rely on a single statistical study. If multiple studies of the same population parameter, using different samples, yield the same or similar results, then one has a compelling body of data. A single study always has a chance of being simply an anomaly, no matter how well the study was conducted.

MEASURES OF CENTRAL TENDENCY

The first and simplest sort of descriptive statistics involves measures of central tendency. This is simply a way of seeing what the aggregate of the data tells us about the data. The three most simple measures of central tendency are the mean, median, and mode. The *mean* is simply the arithmetic average, the *mode* is the item in the sample that appears most often, and the *median* is the item that appears in the middle. Assume you had a set of test scores as follows:

$$65, 74, 84, 84, 89, 91, 93, 99, 100$$

The mode is easy, 84 is the only score that appears more than once. The median is the score in the center, which in this case is 89.

The mean is found by adding the scores and dividing by the number of scores (in this case 9). The formula for that is mean $x = \Sigma x/n$. In this case it would be 86.55. In the preceding formula, x is the mean and n is the number of scores in the sample.

Another important term is *range*. The range is simply the distance from the lowest score to the highest. In our example the highest is 100, the lowest is 65, thus the range is 35.

You will see these four numbers ubiquitously presented in statistical studies. However, what do they really tell us? In this case. the arithmetic mean of the scores was actually about the center of the scores. In our case, all but two of our scores are grouped in a narrow range from 84 to 100. This clustering means that our measures of central tendency probably tell us a lot about our data. But what about situations with much more variety in the numbers? In such cases, the mean may not tell us much about the actual data. This leads to other measures we can do, which can indicate just how accurate the mean is. The standard deviation is a measurement that will tell you this. To quote a popular statistics website:[2]

> The **standard deviation** is a metric that describes how closely the various values are clustered around the mean in a set of data. When the individual values are rather tightly clustered together and the bell-shaped curve is steep, the standard deviation is small. When the values are spread apart and the bell curve is relatively flat, this will produce a relatively large standard deviation.

The standard deviation (denoted *s*) is the square root of the sum of the variance divided by the number of elements. Put in simpler terms, you take each item in the sample, and see how far it varies from the mean. You then square that value and divide it by the number in the sample. Take the square root of that number (which is called the variance) and you have the standard deviation. This can be seen in equation 2.1.

$$ s = \sqrt{\frac{\sum_{n-1}^{n}(x - \bar{x})^2}{n-1}} \qquad \text{(eq. 2.1)}$$

In equation 2.1, *x* is the current value, $\bar{x}$ is the mean, and *n* is the number of values in the sample. In our example we would take each item in the sample minus the mean of 86.5, square that difference, and total the results, like this:

$$(65-86.5)^2 + (74-86.5)^2 + (84-86.5)^2 + (84-86.5)^2 + (89-86.5)^2$$
$$+ (91-86.5)^2 + (93-86.5)^2 + (99-86.5)^2 + (100-86.5)^2$$

which is equal to:

$$462.25 + 156.25 + 6.25 + 6.25 + 6.25 + 20.25 + 42.25 + 156.25 + 182.25$$
$$= 882$$

Now divide that by *n* − 1 (*n* = 9, so divide by 8) and you get 110.25, which is the variance. The square root of the variance, in this case, 10.5, is the standard deviation. That means, in plain English, that on average, the various scores were about 10.5 units from the mean.

As you can see, standard deviation and variance are simply arithmetical computations done to compare the individual items in the sample to the mean. They give you some idea about the data. If there is a small standard deviation, that indicates that values were clustered near the mean and that the mean is representative of the sample.

These are not the only means of measuring central tendency, but they are the most commonly used. Virtually all statistical studies will report mean, median, mode, range, standard deviation, and variance.

CORRELATION

After a study, you have may have two variables, let us call them *x* and *y*. The question is what degree of a correlation do they have? What is the relationship between the two variables? There are a few statistical methods for calculating this. One way to view the relationship between two variables are:

The *Pearson Product-Moment Correlation Coefficient* (r), or correlation coefficient for short *is a measure of the degree of linear relationship between two variables*, usually labeled X and Y. While in regression the emphasis is on predicting one variable from the other, in correlation the emphasis is on the degree to which a linear model may describe the relationship between two variables.

A more clear and concise definition can be found at the BMJ (a medical journal) website:[3] "The word correlation is used in everyday life to denote some form of association. We might say that we have noticed a correlation between foggy days and attacks of wheeziness. However, in statistical terms we use correlation to denote association between two quantitative variables."

One valuable way to calculate this is via Pearson's correlation coefficient, usually denoted with a lower case r. The formula for this is shown in equation 2.2.

$$r = \frac{\sum(x_i - \bar{x})(y_i - \bar{y})}{\sqrt{\sum(x_i - \bar{x})^2 \sum(y_i - \bar{y})^2}}$$

(eq. 2.2)

Equation 2.2 can be summarized with the following steps:

1. Take each value of x and multiply it by each value of y.
2. Multiply n times the mean of x and the mean of y.
3. Subtract the second number from the first and you have the numerator of our equation.
4. Now multiply $n - 1$ times the standard deviation of x and the standard deviation of y, and you now have the numerator.
5. Do the division and you have Pearson's correlation coefficient.

To illustrate this further, let us work out an example. Assume you have x variable of years of postsecondary education, and y variable as annual income in tens of thousands. We will use a small sample to make the math simpler:

Years of post-secondary education	Annual salary in 10's of thousands
2	4
3	4
4	5
4	7
8	10

Now follow the previously described steps:

1. $\Sigma xy = 148$
2. mean $x = 4.2$, mean $y = 6$, $n = 5$, so $6 * 5 * 4.2 = 126$
3. $148 - 126 = 22$
4. The standard deviation of x is 2.28, the standard deviation of y is 2.54, and $n - 1 = 4$, which yields $4 * 2.28 * 2.54 = 23.164$
5. $22/23.164 = .949$ as the Pearson's correlation coefficient.

Now we have calculated the value of r, but what does it mean? Values of r will always be between -1.0 and positive 1.0. A -1.0 would mean a perfect negative correlation, whereas $+1.0$ would indicate a perfect positive correlation. Thus, a value of .949 indicates a very strong positive correlation between years of postsecondary education and annual income.

In our limited sample we found a high positive correlation, but how significant is this finding? Answering the question of significance is where the T test comes in, and fortunately the arithmetic operations for it are much simpler. There are actually several variations of the T-test. The formula for the student's T-test is given in equation 2.3.

$$ t = \frac{\bar{x} - \mu}{s / \sqrt{n}} \qquad \text{(eq. 2.3)} $$

In this formula we have n as the sample size, $\bar{x}$ is the sample mean, μ is the population mean, and s is the variance. This value is sometimes referred to as the coefficient of determination.

In our scenario we have $n = 5$ and $r = .949$, so we have $\sqrt{(3/.099399)}$, which is 5.493, which we multiply by r to get 5.213 for T. This T-score gives us some idea of how significant our correlation coefficient is.

P-Value

Another way to determine if the relationship between our variables is statistically significant is the P-value. According to the Math World[4] website, a p-value is "The probability that a variate would assume a value greater than or equal to the observed value strictly by chance." AstroStat explains p-values this way:[5] "Each statistical test has an associated null hypothesis, the p-value is the probability that your sample could have been drawn from the population(s) being tested (or that a more improbable sample could be drawn) given the assumption that the null hypothesis is true. A p-value of .05, for example, indicates that you would have only a 5% chance of drawing the sample being tested if the null hypothesis was actually true.

Null hypotheses are typically statements of no difference or effect. A p-value close to zero signals that your null hypothesis is false, and typically that a difference is very likely to exist. Large p-values closer to 1 imply that there is no detectable difference for the sample size used. A p-value of 0.05 is a typical threshold used in industry to evaluate the null hypothesis. In more critical industries (healthcare, etc.) a more stringent, lower p-value may be applied."

Z-Test

The z-test is used to make inferences about data. Its formula is shown in equation 2.4.

$$ z = \frac{\mu - x}{\sigma} \qquad \text{(eq. 2.4)} $$

- σ (the standard deviation of the population)
- μ (the mean of the population)
- x (the mean of the sample)

The z-score is equal to the mean of the population minus the mean of the sample divided by the standard deviation of the population. The z-test is often used when you wish to compare one sample mean against one population mean. But what does it actually tell us? To begin with, you need to select a threshold of validity for your statistical study. The z-score converts your raw data (your sample mean) into a standardized score that can be used (along with your preselected threshold of rejection) to determine if you should reject the null hypothesis or not. Z-scores always have a mean of zero and a standard deviation of 1.

Most statistics textbooks will have a z-score chart. You can find what percentage of sample items should appear above or below your z-score. You now compare that to your preselected rejection level. If you previously decided that a 5% level would cause you to reject the null hypothesis and your z-score indicates 11%, then you will reject the null hypothesis. So, the z-score is essential in helping you to decide whether or not to reject the null hypothesis.

OUTLIERS

A significant problem for any statistical study is the existence of outliers. An outlier is a value that lies far outside where most other values lie. For example, if a sample of high school athletes all have heights ranging from 70 inches to 75 inches, except for one who has a height of 81 inches, that one will skew all statistics that you generate. His height will make the mean much higher and the standard deviation much wider. Statistics generated by including that data point in the sample will not accurately reflect the sample. So, what can be done?

One solution to outliers is to exclude them from the data set. It is common to exclude data points that are 2 or more standard deviations from the mean. Therefore, if the mean is 74 inches, and the standard deviation is 3 inches, then any height that is more than 80 inches or less than 68 inches is excluded from the data set. There are, of course, varying opinions on just how far from the mean constitutes a standard deviation, and whether or not they should be excluded at all. If you are conducting a study and elect to exclude outliers, it is a good idea to indicate that in your study, and what method you chose to exclude outliers.

T-Test

T-tests are often used when the research needs to determine if there is a significant difference between two population means. The general formula is shown in equation 2.5.

$$t = \frac{(\Sigma D)/N}{\sqrt{\dfrac{\Sigma D^2 - \left(\dfrac{(\Sigma D)^2}{N}\right)}{(N-1)(N)}}} \qquad \text{(eq. 2.5)}$$

ΣD is the sum of the differences
ΣD^2 is the sum of the squared differences
$\Sigma D)^2$ is the sum of the differences squared

Equation 2.5 only shows a basic t-test. There are many variations such as a one-sample t-test, a Student's t-test, Welch's t-test, and others. T-tests are actually built into the Python programming language with the library scipy, which will be discussed in detail later in this book. Usually when selecting a t-test, and deciding which t-test to use, the place to start is with questions:

Single sample t: we have only 1 group; want to test against a hypothetical mean.
Independent samples t: we have 2 means, 2 groups; no relation between groups, e.g., people randomly assigned to a single group.
Dependent t: we have two means. Either same people in both groups, or people are related, e.g., left hand-right hand, hospital patient and doctor.

For the t distribution, degrees of freedom will always be a simple function of the sample size, e.g., $(n-1)$. One way of thinking about degrees of freedom is that if we know the total or mean, and all but one score, the last $(n-1)$ score is not free to vary. It is fixed by the other scores. $4 + 3 + 2 + X = 10$. $X = 1$.

LINEAR REGRESSION

As was briefly mentioned in chapter 1, linear regression explores the relationship between a variable and one or more explanatory variables. If there is one explanatory variable, then the process is *simple linear regression*. When there are multiple explanatory variables, then the process is *multiple linear regression*. While there are a variety of techniques for linear regression, they are all based on using linear predictor functions whose unknown parameters are estimated from the data. This allows the function to model the relationships between variables.

For predictive analytics, linear regression is often used to fit a model of the observed data from the response and explanatory variables. This will allow the evaluation of the model to determine the efficacy of its predictive power.

Let us begin by examining linear regression in a general manner. Given some dataset $[y_i, x_1, x_2, \ldots x_n]$ linear regression assumes that the relationship between the dependent variable y and the vector of regressors x is linear. The relationship is modeled using a random variable that adds "noise" to the relationship between the regressors and the dependent variable. This random variable is often called an error variable or disturbance term and is symbolized by ε. The model takes the form shown in equation 2.6:

$$y_i = \beta_0 + \beta_1 x_{i1} + \cdots + \beta_p x_{ip} + \varepsilon_i = \mathbf{x}_i^\top \beta + \varepsilon_i, \quad i = 1,\ldots,n, \qquad \text{(eq. 2.6)}$$

In equation 2.5, the x values are the dataset, the y is the dependent variable, the T superscript is a matrix transpose (recall transpose operation from chapter 1), and $x_i^T \beta$ is the inner product between vectors x and β.

There are quite a few different ways of using linear regression and different formulas. The forgoing discussion was simply a general review of the concept of linear regression.

ADDITIONAL STATISTICS

The preceding portion of this chapter will give you a very general, basic understanding of statistics. There are many more statistical tools that you may find of use in machine learning. A few of these will be briefly introduced here. The purpose of this section is to introduce you to topics you may wish to explore after this chapter.

ANOVA

ANOVA, or analysis of variation, is actually a family of statistical tests. We will focus on just one. The one-way analysis of variance (ANOVA) is used to determine whether there are any statistically significant differences between the means of two or more independent groups. The one-way ANOVA compares the means between the groups and determines if the means are statistically different from each other. This test uses a linear model; there are several different linear models used: one that is commonly used to test for effects is shown in equation 2.7.

$$y_{i,j} = \mu + \tau_j + \varepsilon_{i,j}$$
(eq. 2.7)

In equation 1, the i is the index of experimental units, j is the index over treatment groups, $y_{i,j}$ are observations, μ is the mean of the observations (μ_j is the mean of the jth treatment group, μ by itself is the mean of all observations), T_j is the jth treatment effect (a deviation from the mean).

The Multivariate Analysis of Variance (MANOVA) is used when there are two or more dependent variables. It is a variation of ANOVA. MANOVA uses covariance between variables. Where sums of squares appear in univariate analysis of variance, in multivariate analysis of variance, certain positive-definite matrices appear. It is an extension of ANOVA in which main effects and interactions are assessed on a combination of dependent variables. MANOVA tests whether mean differences among groups on a combination of dependent variables is likely to occur by chance.

THE KRUSKAL-WALLIS

The Kruskal-Wallis test is a nonparametric test that compares two or more independent samples. The Kruskal-Wallis test is essentially a nonparametric version of the ANOVA test. Unlike the ANOVA, the Kruskal-Wallis test does not assume normal distribution of the data set. This test is often used to test correlation between two or more data sets that may not even have the same number of elements. The formula for the Kruskal-Wallis test is given in equation 2.8

$$H = (N-1)\frac{\sum_{i=1}^{g} n_i \left(\overline{r}_{i\cdot} - \overline{r}\right)^2}{\sum_{i=1}^{g}\sum_{j=1}^{n_i} \left(r_{ij} - \overline{r}\right)^2}$$
(eq. 2.8)

In equation 2.6, the N is the total number of observations across all groups, g is the number of groups, n_i is the number of observations for group i, r_{ij} is the ran of observation j from group i. Other elements are indeed equations themselves and are shown in equations 2.9 and 2.10.

$$\bar{r}_{i\cdot} = \frac{\sum_{j=1}^{n_i} r_{ij}}{n_i}$$ (eq. 2.9)

$$\bar{r} = \frac{1}{2}$$ (eq. 2.10)

Equation 2.9 provides the average rank of all observations in group i while equation 2.10 is the average of all r_{ij}.

KOLMOGOROV-SMIRNOV

This is a nonparametric test of the equality of continuous data, sometimes called a K-S test or just KS test. It can be a one-sample to two-sample test. This test quantifies the distance between the distribution function of the sample and a reference distribution function. The empirical distribution function F for n ordered observations X is defined as shown in equation 11.

$$F_n(x) = \frac{1}{n} \sum_{i=1}^{n} I_{[-\infty,x]}(X_i)$$ (eq. 2.11)

This is a bit more of an advanced statistical test, and is often not provided in introductory statistics texts. It is given here for those readers interested in expanding beyond what is provided in this chapter.

STATISTICAL ERRORS

Obviously statistical errors can and do occur. Generally, such errors can be classified as either type I or type II errors.[6] A type I error occurs when the null hypothesis is rejected when in fact it is true. A type II error occurs when the null hypothesis is accepted when in fact it is false. The following table illustrates this:

TABLE 2.1
Statistical Errors

Truth	Decision	
	Reject $H0$	Accept $H0$
$H0$ true	Type I Error	Correct Decision
$H1$ true	Correct Decision	Type II Error

Both type I and type II errors should be avoided. Being aware of the possible errors is critical.

POWER OF A TEST

The power of a statistical hypothesis test measures the test's ability to reject the null hypothesis when it is actually false—that is, to make a correct decision. In other words, the power of a hypothesis test is the probability of not committing a type II error. It is calculated by subtracting the probability of a type II error from 1, usually expressed as:

$$Power = 1 - P(type\ II\ error)$$

The maximum power a test can have is 1, the minimum is 0. The ideal situation is to have high power, close to 1.

BASIC PROBABILITY

In the introduction to this chapter, the concept of probability was introduced. A technical definition might be: probability is a measure of uncertainty. The probability of event A is a mathematical measure of the likelihood of the event's occurring. Also recall that the probability of an event must lie between zero and one. There are some additional probability rules that need to be discussed.

WHAT IS PROBABILITY?

The first task is to define what probability is. In simplest terms it is a ratio between the number of outcomes of interest divided by the number of possible outcomes. For example, if I have a deck of cards consisting of 52 cards, made up of 4 suits of 13 cards each, the probability of pulling a card of a given suit is 13/52 or ¼ = .25. Probabilities are always between zero and one. Zero indicates absolutely no chance of an event occurring. If one removes all 13 clubs from a deck, the odds of then pulling a club are zero. A probability of 1.0 indicates the event is certain. If one removes all the cards except for hearts, then the probability of drawing a heart is 1.0.

Basic Set Theory

Probability often uses set theory; therefore, a basic understanding of the essentials of set theory is necessary before we can continue. Let us begin by stating that a set is simply a collection of elements. An empty set is one containing no elements, and the universal set is the set containing all elements in a given context denoted by S. The compliment of a set is the set containing all the members of the universe set that are not in set A. This is denoted by a capital A with a bar over it or by A^c.

As with much of mathematics, terminology and notation is critical. So, let us begin our study in a similar fashion, building from simple concepts to more complex. The

most simple I can think of is defining an element of a set. We say that x is a member
of set A. This can be denoted as

$$x \in A$$

Sets are often listed in brackets. For example, the set of all odd integers < 10 would
be shown as follows:

$$A = \{1, 3, 5, 7, 9\}$$

And a member of that set would be denoted as follows:

$$3 \in A$$

Negation can be symbolized by a line through a symbol. For example:

$$2 \notin A$$

2 is not an element of set A.

If a set is not ending, you can denote that with ellipses. For example, the set of all
odd numbers (not just those less than 10) can be denoted

$$A = \{1, 3, 5, 7, 9, 11, \ldots\}$$

You can also denote a set using an equation or formula that defines membership in
that set.

Sets can be related to each other; the most common relationships are briefly
described here:

Union: If you have two sets A and B, elements that are a member of A, B or both
represent the union of A and B, symbolized as: $A \cup B$.

Intersection: If you have two sets A and B, elements that are in both A and B are
the intersection of sets A and B, symbolized as $A \cap B$. If the intersection of
sets A and B is empty (i.e., the two sets have no elements in common) then
the two sets are said to be disjoint.

Difference: If you have two sets A and B, elements that are in one set, but not
both, are the difference between A and B. This is denoted as $A \setminus B$.

Compliment: Set B is the compliment of set A if B has no elements that are also
in A. This is symbolized as $B = A^c$.

Double Compliment: the compliment of a set's compliment is that set. In other
words, the compliment of A^c is A. That may seem odd at first read, but reflect on the
definition of the compliment of a set for just a moment. The compliment of a set has
no elements in that set. So, it stands to reason that to be the compliment of the compliment of a set, you would have to have all elements within the set.

These are basic set relationships. Now a few facts about sets:

Order is irrelevant: {1, 2, 3} is the same as {3, 2, 1}, or {3, 1, 2}, or {2, 1, 3}.

Subsets: Set A could be a subset of Set B. For example, if set A is the set of all odd numbers < 10 and set B is the set of all odd numbers < 100, then set A is a subset of set B. This is symbolized as: $A \subseteq B$.

Power set: As you have seen, sets may have subsets. Let us consider set A as all integers less than 10. Set B is a subset; it is all prime numbers less than 10. Set C is a subset of A; it is all odd numbers < 10. Set D is a subset of A; it is all even numbers less than 10. We could continue this exercise making arbitrary subsets such as $E = \{4, 7\}$, $F = \{1, 2, 3\}$, etc. The set of all subsets for a given set is called the power set for that set.

Sets also have properties that govern the interaction between sets. The most important of these properties are listed here:

Commutative Law: The intersection of set A with set B is equal to the intersection of set B with Set A. The same is true for unions. Put another way, when considering intersections and unions of sets, the order the sets are presented in is irrelevant. That is symbolized as seen here:

$$(a)\ A \cap B = B \cap A$$
$$(b)\ A \cup B = B \cup A$$

Associative Law: Basically, if you have three sets and the relationships between all three are all unions or all intersections, then the order does not matter. This is symbolized as shown here:

$$(a)\ (A \cap B) \cap C = A \cap (B \cap C)$$
$$(b)\ (A \cup B) \cup C = A \cup (B \cup C)$$

Distributive Law: The distributive law is a bit different than the associative, and order does not matter. The union of set A with the intersection of B and C is the same as taking the union of A and B intersected with the union of A and C. This is symbolized as you see here:

$$(a)\ A \cup (B \cap C) = (A \cup B) \cap (A \cup C)$$
$$(b)\ A \cap (B \cup C) = (A \cap B) \cup (A \cap C)$$

De Morgan's Laws: These govern issues with unions and intersections and the compliments thereof. These are more complex than the previously discussed properties. Essentially, the compliment of the intersection of set A and set B is the union of the compliment of A and the compliment of B. The symbolism of De Morgan's Laws are shown here:

$$(a)\ (A \cap B)^c = A^c \cup B^c$$
$$(b)\ (A \cup B)^c = A^c \cap B^c$$

These are the basic elements of set theory. You should be familiar with them before proceeding.

BASIC PROBABILITY RULES

The following is a brief list of basic probability rules.

- The probability of any event will be between zero and one, $0 \leq P \leq 1.0$.
- Probability of the complement of an event (remember that set theory plays a role in probability) is equal to 1– probability of the event. Or put another way: $P(\overline{A}) = 1 - P(A)$. What this means is that if the probability of a given event A is .45, then its compliment is 1 – .45 or .55
- Rule of unions: The probability of a union of events is the probability of event A plus the probability of event B minus the probability of their intersection (or joint probability).
- Joint probability of independent events: This is simply the probability of event A multiplied by the probability of event B.

$$P(A \text{ and } B) = P(A) * P(B)$$

- As an example, if two events are independent and event A has a probability of .45 and event B has a probability of .85, then the probability of both events occurring is .45 * .85 = .3825
- For two mutually exclusive events, the probability of their union is simply the probability of event A + the probability of event B. $P(A \cup B) = P(A) + P(B)$.

These basic rules are important to probability and should be committed to memory by any student who wishes to successfully study probability.

CONDITIONAL PROBABILITY

Conditional probability refers to the likelihood of an event occurring given some other event occurring. The likelihood of event A occurring, given event B has occurred, is equal to the probability of the intersection of event A and B divided by the probability of event B. This rule obviously is not referring to situations where event B must follow A, but where event A can lead to event B. For example, if it is cold, there is a certain probability that I will wear a jacket, this is conditional probability. The probability of my wearing a jacket is conditioned upon the temperature.

INDEPENDENT EVENTS

Independent events are events whose probability has no relationship at all. Put another way, two events are independent if the following are true (and conversely the following statements are true if the two events are independent):

- $P(A \mid B) = P(A)$
- $P(B \mid A) = P(B)$

The intersection of two or more independent events is just the product of their separate probabilities.

BAYES THEOREM

Thomas Bayes was a clergyman in the 18th century whose work has been very influential in statistics and probability. Bayes' theorem is a mathematical formula used for calculating conditional probabilities. It is the basis of *Bayesian* approaches to epistemology, statistics, and inductive logic. The theorem's central insight is simply that a hypothesis is confirmed by data that its truth renders. The probability of a hypothesis H, conditional on a given body of data E, is the ratio of the unconditional probability of the conjunction of the hypothesis with the data to the unconditional probability of the data alone. Bayes' theorem is defined as:

The probability of H conditional on E is defined as $\mathbf{P}_E(H) = \mathbf{P}(H \& E)/\mathbf{P}(E)$, provided that both terms of this ratio exist and $\mathbf{P}(E) > 0$.

That definition may seem a bit convoluted to the novice; it may help elucidating this concept to examine a hypothetical example. Assume a randomly chosen American who was alive on January 1, 2000. According to the United States Center for Disease Control, roughly 2.4 million of the 275 million Americans alive on that date died during the 2000 calendar year. Among the approximately 16.6 million senior citizens (age 75 or greater) about 1.36 million died. Now consider our hypothesis that our subject died during the 2000 calendar year. Essentially, the unconditional probability of the hypothesis that our subject died during 2000, H, is just the population-wide mortality rate $\mathbf{P}(H) = 2.4M/275M = 0.00873$. To find the probability of J. Doe's death conditional on the information, E, that he was a senior citizen, we must divide the sample space into two different areas (though more than two can be used). We divide the probability that he or she was a senior who died, $P(H \& E) = 1.36M/275M = 0.00495$, by the probability that he or she was a senior citizen, $P(E) = 16.6M/275M = 0.06036$. Thus, the probability of our subject's death given that he was a senior citizen is $P_E(H) = P(H \& E)/P(E) = 0.00495/0.06036 = 0.082$. Notice how the size of the *total* population factors out of this equation, so that $P_E(H)$ is just the proportion of seniors who died. Bayes' theory allows us to work with conditional probabilities more efficiently. Expressed as a formula, Bayes' theory is

$$P_E(H) = [P(H)/P(E)] \, P_H(E)$$

A Bayesian examination of conditional probability allows one to evaluate the predictive value of certain factors. Statisticians refer to the inverse probability $P_H(E)$ as the "likelihood" of H on E. It expresses the degree to which the hypothesis *predicts* the data given the background information codified in the probability P. In the example discussed previously, the condition that our subject died during 2000 is a fairly strong predictor of senior citizenship. Indeed, the equation $P_H(E) = 0.57$ tells us that 57% of the total deaths occurred among seniors that year. Bayes' theorem lets us use this information to compute the probability of our subject dying given that he was a senior citizen. We do this by multiplying the "prediction term" $P_H(E)$ by the ratio of the total number of deaths in the population to the number of senior citizens in the population, $P(H)/P(E) = 2.4M/16.6M = 0.144$. The result is $P_E(H) = 0.57 \times 0.144 = 0.082$, just as expected.

Bayes' theorem is of value in calculating conditional probabilities, because inverse probabilities are typically both easier to ascertain and less subjective than direct probabilities. People with different views about the unconditional probabilities of E and H often disagree about E's value as an indicator of H. Even so, they can agree about the degree to which the hypothesis predicts the data if they know any of the following intersubjectively available facts: (a) E's *objective* probability given H, (b) the frequency with which events like E will occur if H is true, or (c) the fact that H logically entails E. Scientists often design experiments so that likelihoods can be known in one of these "objective" ways. Bayes' theorem then ensures that any dispute about the significance of the experimental results can be traced to "subjective" disagreements about the unconditional probabilities of H and E.

When both $P_H(E)$ and $P_{\sim H}(E)$ are known, an experimenter need not even know E's probability to determine a value for $P_E(H)$ using Bayes' theorem.

Bayes' theorem (second form):

$$P_E(H) = P(H)P_H(E)/[P(H)P_H(E) + P(\sim H)P_{\sim H}(E)]$$

In this form, Bayes' theorem is particularly useful for inferring causes from their effects, since it is often fairly easy to discern the probability of an effect given the presence or absence of a putative cause.

SPECIAL FORMS OF BAYES' THEOREM

Bayes' theorem can be expressed in several different forms. Each is useful for different purposes. One version employs what is often called the *relevance quotient* or *probability ratio*. This is the factor $PR(H, E) = P_E(H)/P(H)$, by which H's unconditional probability must be multiplied to get its probability conditional on E. Bayes' theorem is equivalent to a simple symmetry principle for probability ratios.

Probability ratio rule: $PR(H, E) = PR(E, H)$

The term "Probabiliy ratio rule" provides one measure of the degree to which H *predicts E. If we think of $P(E)$ as expressing the "baseline" predictability of E given the background information codified in P, and of $P_H(E)$ as E's predictability when H is added to this background, then $PR(E, H)$ captures the degree to which knowing H makes E more or less predictable relative to the baseline: $PR(E, H) = 0$ means that H categorically predicts $\sim E$; $PR(E, H) = 1$ means that adding H does not alter the baseline prediction at all; $PR(E, H) = 1/P(E)$ means that H categorically predicts E.

Another commonly encountered form of Bayes' theorem is referred to as *Odds Rule*. In popular terminology, the "odds" of a hypothesis is its probability divided by the probability of its negation: $O(H) = P(H)/P(\sim H)$. So, for example, a football team, whose odds of winning a particular match are eight to two, has an 8/10 chance of winning and a 2/10 chance of losing. Contrary to popular thought, probability and odds are not necessarily the same.

Odds ratio rule: $OR(H, E) = P_H(E)/P_{\sim H}(E)$

SUMMARY

This chapter has provided a brief overview of statistics. A general working knowledge of statistics is important to machine learning. You should ensure you are familiar with all the concepts introduced in this chapter. Keep in mind that the material in this chapter is just a brief introduction of the essentials. If you intend to work in data science and machine learning, you will certainly need to have a deeper knowledge of statistics than what is presented here. For those who would like to go beyond the material in this chapter, the following resources may be of use to you:

Cramér, H. 2016. *Mathematical methods of statistics*. Princeton University Press.

Peck, R., Olsen, C. and Devore, J. L. 2015. *Introduction to statistics and data analysis*. Cengage Learning.

TEST YOUR SKILLS

1. A researcher is studying patients who have been diagnosed with traumatic brain injury. The group of subjects in his or her specific study are an example of a _____.

   A. Parameter

   B. Statistic

   C. Population

   D. Sample

2. Five-point Likert scales (strongly disagree, disagree, neutral, agree, strongly agree) are frequently used to measure motivations and attitudes. A Likert scale is a:

   A. Nominal variable.

   B. Ordinal variable.

   C. Interval variable.

   D. All of the above options (A, B, and C)

3. A physician surveys neurological patients to determine if the patient has had an MRI, CAT, or PET scan. What type of data is this?

   A. Nominal

   B. Ordinal

   C. Interval

   D. Ratio

4. What is the mean of the following set: 4, 2, 4.5, 4, 5, 7

   A. 4

   B. 4.41

   C. 5

   D. 4.8

5. What is the standard deviation of the following set: 4, 7, 1, 5, 4, 3

   A. 1.8

   B. 3.6

   C. 4

   D. 3.3

6. Which of the following is best used to determine if the relationship of variables is statistically significant?

   a. T-Test

   b. Z-Test

   c. P-Value

   d. Kruskal-Wallis Test

7. The _____ is used to determine whether there are any statistically significant differences between the means of two or more independent groups.

   a. ANOVA

   b. T-Test

   c. Kruskal-Wallis Test

   d. P-Value

8. _____ explores the relationship between a variable and one or more explanatory variables.

   a. Kruskal-Wallis Test

   b. Z-Test

   c. Linear Regression

   d. T-Test

NOTES

1. https://www.cornellcollege.edu/library/ctl/qr/PDFs/StatisticsGlossary.pdf
2. https://www.investopedia.com/terms/s/standarddeviation.asp
3. https://www.bmj.com/about-bmj/resources-readers/publications/statistics-square-one/11-correlation-and-regression
4. https://mathworld.wolfram.com/Significance.html
5. http://voi.iucaa.in/help/astrostat/
6. https://www.ncbi.nlm.nih.gov/pmc/articles/PMC2996198/

3 Introduction to Python Programming

INTRODUCTION

Python is a popular scripting language. It is open-source, meaning the actual source code for Python itself is available and there is no charge to use Python. It was invented in the early 1990s by Guido van Rossum. The name is not derived from the Python snake, but rather the British comedy team, Monty Python. The language structure is managed by the Python Software Foundation www.python.org. The current version of Python is 3.0, which was released in 2008.

Why is Python the focus of the programming in this book? It is certainly true that one can do machine learning programming in many other programming languages. However, Python has several advantages. It is open-source, as was discussed in the previous paragraph. Furthermore, Python is widely used in machine learning and data science. You will even be able to find numerous code samples on the internet that you can modify for your needs. Python is also fairly simple to use, even if you are new to programming. Finally, there are Python packages for just about any task you may wish to accomplish. The Python Package Index (PyPI) lists 130,000 packages.

FUNDAMENTAL PYTHON PROGRAMMING

This section is for those readers who have very limited programming experience. If you have experience in programming, particularly in something C-based (C, C++, Java, etc.) then you may wish to just skim this section to at least get some of the Python-specific items. However, if you are new to programming, then it is important that you read this section and ensure you fully understand the materials before proceeding with the rest of the book.

VARIABLES AND STATEMENTS

In general, all programs handle data of some sort. Regardless of the purpose of the program, it must process data. Data must be temporarily stored in the program. This is accomplished via variables. A variable is simply a place in memory set aside to hold data of a particular type. These are called variables because their value or content can change or vary. When you create a variable, you are essentially allocating a small piece of memory for storage purposes. The name you give the variable is simply a label for that address in memory. Python actually creates a variable for you the first time you reference a variable. So, if you have the following:

$$X = 10$$

You have just allocated four bytes of memory (the amount used by integers), and you are using the variable x to refer to those four bytes of memory. You are also making x an integer type, by virtue of the fact that you first set it equal to an integer. Now whenever you reference x in your code, you are actually referencing a specific address in memory. Table 3.1 lists and describes the basic data types available in Python

Once you have variables, the next building block is a statement. A statement is simply a single line of code that performs some action. That action might be to declare a variable, add two numbers, compare two values, or just about anything at all. All statements/expressions end with a semicolon. This is true in many programming languages including Java, C, and C#. A few example statements could help clarify this for you:

```
Print("Hello World")
acctnum = 555555
acctnum = acctnum + 5
```

Each statement performs a different action, but it does perform some action. Unlike C, C++, and Java, you do not need to end a statement with a semicolon. In many programming languages, the term statement and expression are used interchangeably.

Now we can move on to functions. A function is a set of statements that are grouped together under a particular name, to perform some specific task. Consider the following, rather simple, function:

```
def my _ function():
print("Hello from a function")
```

TABLE 3.1

Python Data Types

Data Type	Description
str	String or text
int	Represents a four-byte integer
float	Represents a four-byte decimal value such as 3.14
bool	Represents a Boolean value: true/false
byte	Represents a binary byte. There is also a byte array type
complex	Represents complex numbers such as 3y + 4i

This short function teaches us quite a bit about Python programming. First, all functions start with the def statement. That is defining a function. Then is the name of the function. If that function requires you to pass it any information (what we call arguments or parameters to the function) those go in the parentheses. We will see a function that takes in arguments a little later in this section.

Let us now look at another function, then discuss some details about Python programming. The following is a very simple Python script, but by examining every line, you will learn a great deal about Python.

```python
def main():
    #first get the user input
    Celsius = eval(input("What is the Celsius temperature? "))
    Fahrenheit = 9 / 5 * Celsius + 32
    print("The temperature is", Fahrenheit, "degrees Fahrenheit.")
    if Fahrenheit >= 90:
    print("It's really hot out there, be careful!")
    if Fahrenheit <= 30:
        print("Brrrrr. Be sure to dress warmly")
        main()
```

Before we examine this, let us also see how it looks in the most common Python editor, Idle. This is shown in figure 3.1.

Executing this script is shown in figure 3.2.

Now that you have seen this script function, let us return to examining the script itself. From examining this rather elementary Python script, you should be able to learn quite a bit of fundamental Python. The first line is shown starting with the # symbol. In Python, that denotes a comment. Comments are ignored by the computer, and are simply there to provide information for other programmers. When you are new to any programming language, comments can help you learn the language. It is also frequently the case that one will return to code that was written many months in the past, and not recall why certain programming decisions were made. Comments

simpleprogram.py - E:/Projects/publishing/Machine Learning For Neuroscience/simpleprogr... -

File Edit Format Run Options Window Help

```python
def main():
    #first get the user input
    celsius = eval(input("What is the Celsius temperature? "))
    fahrenheit = 9 / 5 * celsius + 32
    print("The temperature is", fahrenheit, "degrees fahrenheit.")
    if fahrenheit >= 90:
        print("It's really hot out there, be careful!")
    if fahrenheit <= 30:
        print("Brrrrr. Be sure to dress warmly")

main()
```

FIGURE 3.1 Example script in Idle.

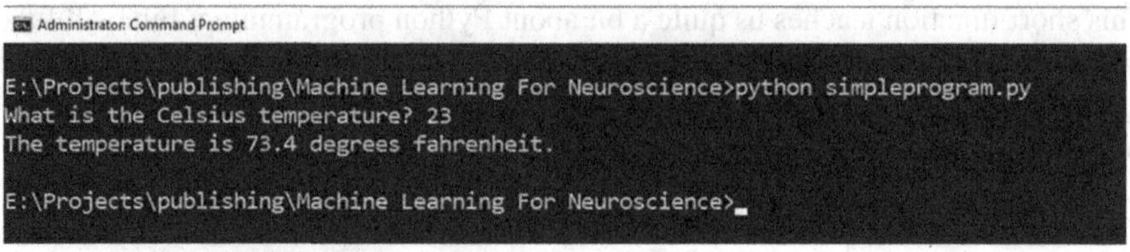

```
Administrator: Command Prompt
E:\Projects\publishing\Machine Learning For Neuroscience>python simpleprogram.py
What is the Celsius temperature? 23
The temperature is 73.4 degrees fahrenheit.

E:\Projects\publishing\Machine Learning For Neuroscience>_
```

FIGURE 3.2   Executing example script.

aid in clarifying this. For these reasons, it is strongly recommended that you use comments liberally.

The next item to note from this code sample is the def main(): this is how functions are created in Python. The def statement means to define. What is being defined is a function with the name immediately following the def statement, in this case main. As you can probably surmise, main is a logical place to begin a program. Another item to note is the indention. Many programming languages (C, C++, Java, C#, etc.) use brackets to define the boundaries of any function, loop, or decision statement. Python uses indention instead. So, pay very close attention to indenting properly. You may also notice that the if statement later in the code uses indention. This is how Python defines a code section.

There are two other statements that are noteworthy. The first gets input from the user. It requests the user enter some value, so it can be converted. Along with that is a print statement. Getting input from a user and producing output are common programming tasks. But there is more to the print statement than might initially occur to you. This print statement actually demonstrates one of the difficulties with Python. That difficulty is that new versions of Python can frequently change how certain things are done, and thus break old code. Printing in version 2.0 of Python was:

```
print "whatever you want to print"
```

Now the parentheses are needed:

```
print ("whatever you want to print")
```

One benefit to using Python is there are so many code samples on the internet. If you just use your favorite search engine and search for "Python script for x," putting whatever you want a script for in place of x, you will likely find numerous sample scripts to download. However, some of these will have been created with an older version of Python and will generate errors when you attempt to use them. Keep that in mind.

When naming variables, you cannot use a Python keyword. These are terms that are set aside to have specific meaning in Python. A list of the more common keywords is given in the following table.

In addition to keywords, you will also encounter conditional statements. This is when you wish to do something, only if some conditions are met. The following table shows the Python conditional statements.

TABLE 3.2

Python Keywords

Keyword	Description
and	A logical operator
break	To break out of a loop
class	To define a class
continue	To continue to the next iteration of a loop
def	To define a function
del	To delete an object
elif	Used in conditional statements, same as else if
else	Used in conditional statements
except	Used with exceptions, what to do when an exception occurs
False	Boolean value, result of comparison operations
finally	Used with exceptions, a block of code that will be executed no matter if there is an exception or not
for	To create a for loop
from	To import specific parts of a module
global	To declare a global variable
if	To make a conditional statement
import	To import a module
not	A logical operator
or	A logical operator
return	To exit a function and return a value
True	Boolean value, result of comparison operations
try	To make a try . . . except statement
while	To create a while loop
with	Used to simplify exception handling

TABLE 3.3

Python Conditional Statements

Python	Meaning
<	Less than
<=	Less than or equal to
==	Equal to
>=	Greater than or equal to
>	Greater than
!=	Not equal to

You will use these later when working with if statements and any conditional statement.

Another common task in Python is retrieving a date. Getting the current date is quite simple and is shown here:

```
import datetime
currdate = datetime.datetime.now()
print(currdate)
```

OBJECT-ORIENTED PROGRAMMING

Python supports object-oriented programming. Therefore, you should be at least basically familiar with object-oriented programming concepts. An *object* is a programming abstraction that groups data with the code that operates on it. All programs contain data of different types. An object simply wraps all that up in a single place, called an object.

In object-oriented programming, there are four concepts that are integral to the entire process of object-oriented programming:

Abstraction is basically the ability to think about concepts in an abstract way. You can create a class for an employee, without having to think about a specific employee. It is abstract and can apply to any employee.

Encapsulation is really the heart of object-oriented programming. This is simply the act of taking the data, and the functions that work on that data, and putting them together in a single class. Think back to our coverage of strings, and the string class. The string class has the data you need to work on (i.e., the particular string in question) as well as the various functions you might use on that data, all wrapped into one class.

Inheritance is a process whereby one class inherits, or gets, the public properties and methods of another class. The classic example is to create a class called animal. This class has properties such as weight, and methods such as move and eat. All animals would share these same properties and methods. When you wish to create a class for, say, a monkey, you then have class monkey inherit from class animal, and it will have the same methods and properties that animal has. This is one way in which object-oriented programming supports code reuse.

Polymorphism literally means "many forms." When you inherit the properties and methods of a class, you need not leave them as you find them. You can alter them in your own class. This will allow you to change the form those methods and properties take.

The term class is one you see a lot in object-oriented programming. It is a template for instantiating objects. Think of it as the blueprint for the objects you will need. It defines the properties and methods of that object.

IDE

One normally uses an Integrated Development Environment (IDE) to write code. Most IDEs will also provide some additional features that can be helpful. For example, IDLE is the IDE that is installed when you install Python. In addition to letting you write scripts, it also has some useful features. We will explore IDLE as well as other IDEs in the following subsections.

IDLE

You may have noticed in the previous code sample (see figure 3.1) that the lines were color-coded. This makes it very easy to note comments, function names, etc., in the code. In addition, the menu bar at the top of IDLE has several tools for you. Let us start with looking at the Options menu, which you can see in figure 3.3.

Showing line numbers is often quite useful. The Configure IDLE will also be useful. The settings allows you to configure IDLE. This is shown in figure 3.4.

These settings allow you to configure the code colors, extensions, fonts, etc., to whatever is most convenient for you. The Run menu option, obviously, allows you to run your script. In this book, the examples will usually be executed from a command line/shell. Also under the Run menu is an option to check the code. This is not meant as a substitute for good programming practices, but can find issues with your code. You can see this menu option in figure 3.5.

Obviously, there are other menu options in IDLE. These will be explored as they come up in the natural progression through machine learning scripts in this book.

OTHER IDEs

There are, of course, other IDEs one can use. Each has its own advantages and disadvantages, as well as proponents and detractors. Python Anywhere is a web-based IDE. There are free versions and paid versions. You can find the website at www. pythonanywhere.com/.

Figure 3.6 shows the landing page for Python Anywhere.

Given the cost for at least some of the options, it is not quite as popular with Python developers. Micro Python is another web-based Python IDE, but this one is free. The website is: https://micropython.org/unicorn/. You can see the landing page for Micro Python in figure 3.7.

Sublime is a free editor that one downloads to one's computer. The website is found at www.sublimetext.com/. The interface is shown in figure 3.8.

FIGURE 3.3 The IDLE options menu.

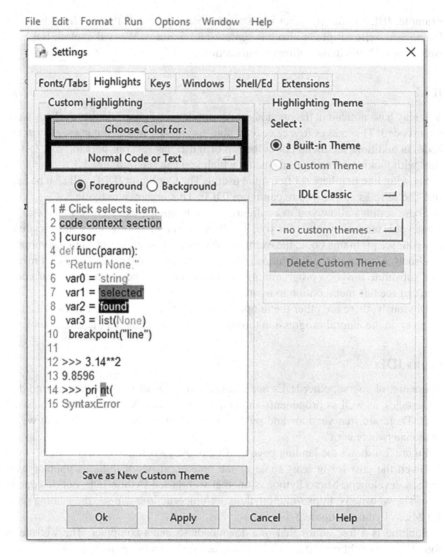

FIGURE 3.4 IDLE settings.

File   Edit   Format   **Run**   Options   Window   Help

Run Module	F5
Run... Customized	Shift+F5
Check Module	Alt+X
Python Shell	

FIGURE 3.5 Run menu.

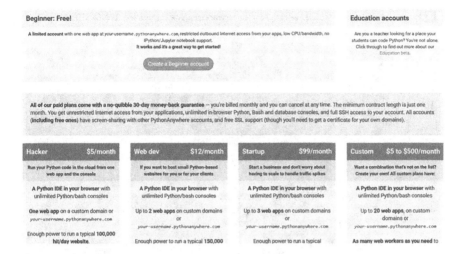

FIGURE 3.6 Python Anywhere.

FIGURE 3.7 Micro Python.

The IDE you use is not a critical issue. You should find one that is comfortable for you and use that one. However, in this book, the IDLE IDE will be used in the code examples you see.

PYTHON TROUBLESHOOTING

There are a number of issues that will arise when using Python. Some of the more common issues are addressed in this section. The objective is to make your use of Python go smoothly. However, no text can account for every issue you may encounter. Fortunately, the Python community is quite helpful. Should you encounter an issue that is not covered in this section, simply searching the internet for the error message will often yield very helpful results.

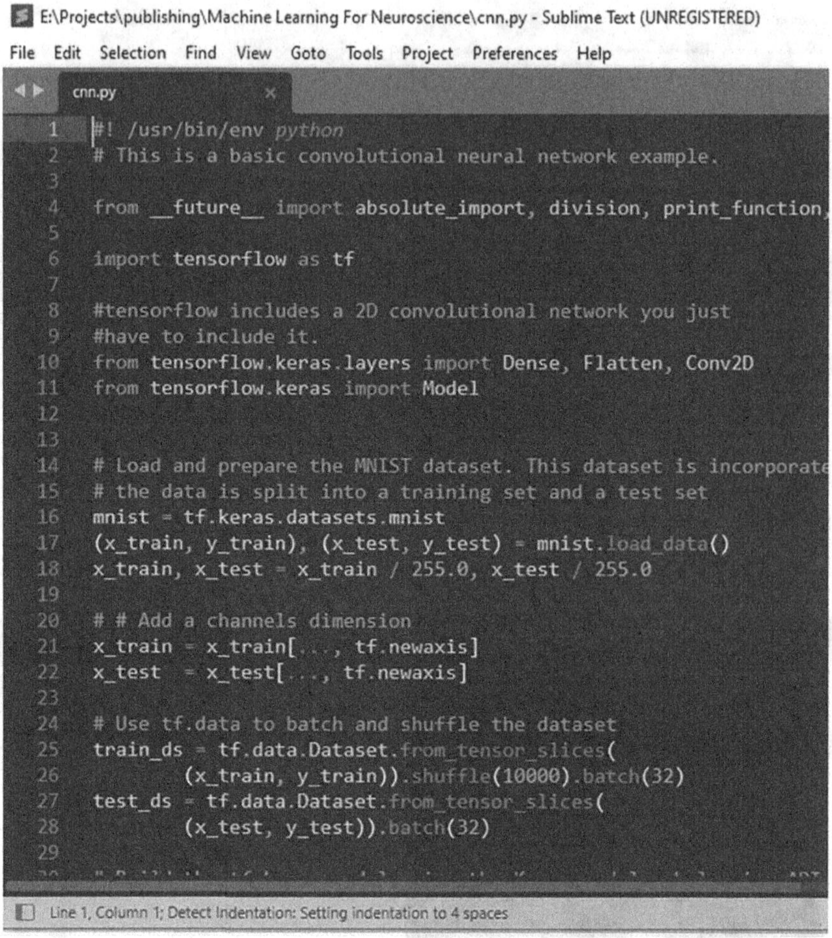

E:\Projects\publishing\Machine Learning For Neuroscience\cnn.py - Sublime Text (UNREGISTERED)

File   Edit   Selection   Find   View   Goto   Tools   Project   Preferences   Help

```
     cnn.py                          x
 1   #! /usr/bin/env python
 2   # This is a basic convolutional neural network example.
 3
 4   from __future__ import absolute_import, division, print_function,
 5
 6   import tensorflow as tf
 7
 8   #tensorflow includes a 2D convolutional network you just
 9   #have to include it.
10   from tensorflow.keras.layers import Dense, Flatten, Conv2D
11   from tensorflow.keras import Model
12
13
14   # Load and prepare the MNIST dataset. This dataset is incorporate
15   # the data is split into a training set and a test set
16   mnist = tf.keras.datasets.mnist
17   (x_train, y_train), (x_test, y_test) = mnist.load_data()
18   x_train, x_test = x_train / 255.0, x_test / 255.0
19
20   # # Add a channels dimension
21   x_train = x_train[..., tf.newaxis]
22   x_test  = x_test[..., tf.newaxis]
23
24   # Use tf.data to batch and shuffle the dataset
25   train_ds = tf.data.Dataset.from_tensor_slices(
26           (x_train, y_train)).shuffle(10000).batch(32)
27   test_ds = tf.data.Dataset.from_tensor_slices(
28           (x_test, y_test)).batch(32)
29
```

Line 1, Column 1; Detect Indentation: Setting indentation to 4 spaces

FIGURE 3.8   Sublime.

The first issue is for Windows users. When using Windows, you need to ensure that Python is in your environment variables. If not, then you can only execute scripts from the directory where the Python.exe is installed. That can be cumbersome. You can see how to set Windows environment variables in figures 3.9 and 3.10.

You can avoid this problem if, during the installation, you make sure you also check to Add Python to environment variables, as shown in figure 3.11.

One of Python's strengths is the huge number of packages available. There are packages for almost any task one can imagine. And we will use several of these frequently when writing machine learning code. However, if you don't have a package your script needs, when you run the script you will get an error message. The obvious thing to do is to attempt to install the package. An example of installing a package is shown in figure 3.12.

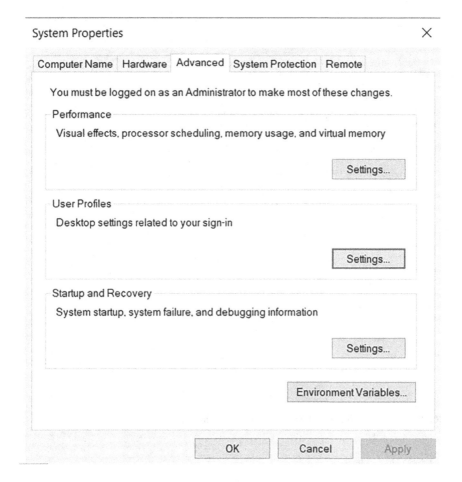

FIGURE 3.9   Setting Windows environment variables step 1.

However, you may find this does not work. You could get an error like this one:

EnvironmentError: [WinError 5] Access is denied:
It is easy to fix the command prompt as Administrator. If in Linux, run as root/superuser. Then attempt the installation. We will be describing several commonly used packages in chapter 4.

GENERAL TIPS TO REMEMBER

As you proceed through the chapters, certain helpful hints will be included. However, there are some general pointers that you need to keep in mind as you go forward. These are issues that, if you are aware of them from the outset, can make your Python programming experience more productive and less frustrating.

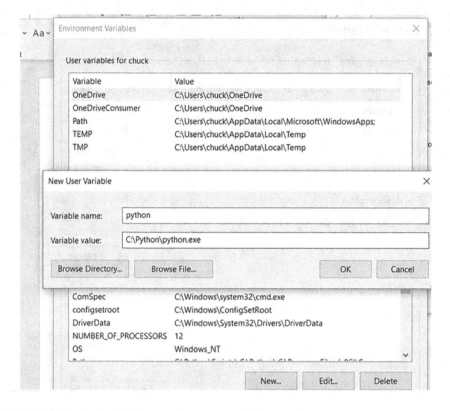

FIGURE 3.10   Setting Windows environment variables step 2.

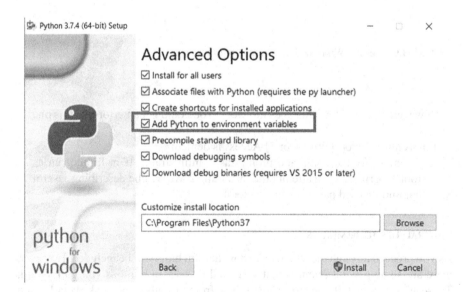

FIGURE 3.11   Windows Python installation.

```
TorBot-dev>pip install PySocks
Collecting PySocks
  Using cached https://files.pythonhosted.org/packages/cd/18/1
02cc70347486e75235a29a6543f002cf758042189cb063ec25334993e36/Py
Socks-1.7.0-py3-none-any.whl
Installing collected packages: PySocks
Successfully installed PySocks-1.7.0
```

FIGURE 3.12   Installing Python packages.

- Indentions mean something in Python. Python does not use brackets or braces, it uses indention.
- Running scripts will be easier if the script's name is all one word (no spaces) and no spaces in the path to the script. That is simply easier for you to type in.
- Be aware of single and double quotes.
- IDLE will start with some header information that may or may not match your version of Python. That may cause problems compiling. Start a new file and there is no header information, and problem solved.

BASIC PROGRAMMING TASKS

After reading the previous sections of this chapter, you should be able to install Python and write basic programs, such as the ubiquitous Hello World. In this final section, we will address common programming tasks. The focus here is not so much on entire scripts, but rather techniques you will frequently use in Python programming.

CONTROL STATEMENTS

There are times when you will want to execute a portion of your code only if some value is present. Or you may wish to execute a section of code a certain number of times. These processes require control statements. The most basic of these is the if statement. If some value is met, then do something. Here is an example:

```
if variableA > 10 then
   print("variableA is greater than 10")
You can also expand this to handle different circumstances such as:
if variableA < 10 then
   print("variableA is less than 10")
elif variableA < 100 then
   print("variableA is between 10 and 100")
else
   print("variableA is greater than 100")
```

If statements are quite common and you will find them in almost all programming and/or scripting languages. As you go further in this book, you will see such state-

---

TABLE 3.4

Python Logical Statements

Operation	Symbol	Example
Equals	==	$1 + 2 == 3$
Not equal	!=	$1 + 2 != 7$
Less than	<	$1 < 2$
Greater than	>	$2 > 1$
Less than or equal to	<=	$1 <= 2$
Greater than or equal	>=	$2 >= 1$

---

ments in action quite frequently. This naturally leads to the variety of logical statements that can be used in these control structures. The following table provides a summary of these.

Another type of control statement is the loop. There are actually several different types of loops. The most common is the for loop. This can be used to loop through an array. The following example should help elucidate this.

```
colors = ["red", "blue", "green"]
for x in colors:
 print(x)
```

This code will loop through the array of colors and print each color to the screen. You can combine this with an if statement to make slightly more clever code, as seen in this example:

```
colors = ["red", "blue", "green"]
for x in colors:
 if x == "blue":
  break
 print(x)
```

This bit of code combines if statements, for loops, and even uses the break statement to break out of the loop. Another loop you will encounter frequently is the while loop. This loop essentially states that while some conditions are true, execute some piece of code. The following rather elementary example should elucidate this concept:

```
i = 1
while i < 6:
    print(i)
 i += 1
```

You will see these basic control statements throughout this book in many programming examples.

WORKING WITH STRINGS

As you can probably surmise, strings are commonly used in any programming language. Strings can be used to represent a person's name, a file path, a domain name, and many other items that are commonly needed in any programming languages. Python provides you a number of techniques to work with strings. Assume you have a string variable named mystring. Table 3.5 has a few examples of what you can do with that variable in Python.

Here is code implementing some of the preceding methods:

```
#first let us create a string
mystring = "machine learning for neuroscience"
#now we can print out various string function results:
print(mystring.capitalize())
print(mystring.isdigit())
print(mystring.islower())
```

When executed, you will see output like in figure 3.13.

The items in the preceding table are just the more common string methods. There are many more, though less commonly used. You can find a comprehensive list at www.w3schools.com/python/python_ref_string.asp.

TABLE 3.5
Python String Functions

mystring.capitalize()	Converts the first character to upper case.
mystring.upper()	Converts a string into upper case.
mystring.lower()	Converts string into lower case.
mystring.isdigit()	Returns true if all characters in the string are digits.
mystring.islower()	Checks if the string is all lower case.
mystring.strip()	Returns a version of the string with leading and trailing spaces removed.
mystring.isnumeric()	Returns true if the string contains all numeric values.
mystring.find('n')	Finds the first instance of the letter "N."
mystring.count('s')	Counts the number of "s" that appear in the string.

```
E:\Projects\publishing\Machine Learning For Neuroscience>python strings.py
Machine learning for neuroscience
False
True
```

FIGURE 3.13   String output.

WORKING WITH FILES

The first step in working with a file is to open it. In Python that is relatively simple, as you can see here:

```
fMyFile = open('mytext','r')
```

You may notice the 'r' and wonder what that is. This means to open the file in read-only mode. The other options are 'w' (write mode), and 'a' (append mode). For example, if you wish to open a file and read the contents onto the screen, you can combine a while loop with the open statement and the print statement:

```
with open('mytext.txt') as fMyFile:
    contents = fMyFile.read()
    print(contents)
```

A SIMPLE PROGRAM

This is a trivial program, but it brings together many things you have learned. It prints to the screen, has if statements, defines a main function, etc. These are all items you will need to do frequently in your code.

```
# converttemp.py
# A program to convert Celsius temperature to Fahrenheit.

def main():
    Celsius = eval(input("What is the Celsius temperature? "))
Fahrenheit = 9 / 5 * Celsius + 32
    print("The temperature is", Fahrenheit, "degrees Fahrenheit.")
    if Fahrenheit >= 90:
        print("It's hot, are you in Texas?")
    if Fahrenheit <= 30:
    print("It is cold")

main()
```

Note that this brief bit of code also has variables named in a way that makes it obvious what they do, and is using comments to explain. The basic programming items you will need are also illustrated. A main function is defined, user input is solicited, output is presented on the screen, and a decision statement (i.e., the if statements) is illustrated. While quite simple, this code sample provides a basic structure you can use in all of your Python programming.

BASIC MATH

You will frequently find the need to utilize various mathematical operations in your Python programming. The basic math operations are shown in table 3.6:

TABLE 3.6
Python Math Operations

Operation	Symbol	Example
Addition	+	4 + 2
Subtraction	-	4–2
Multiplication	*	4 * 2
Division	/	4/2
Exponent	**	4**2
Modulo	%	4 % 2

TABLE 3.7
Python Math Functions

Operation	Python command
Absolute value	abs(somevariable)
Cosine	cos(someangle)
Sine	sin (someangle)
Natural Logarithm	log(somevariable)
Logarithm based 10	log10(somevariable)
Round to the nearest whole number	round(somevariable)
Square root of some number	sqrt(somevariable)

These functions will suffice for elementary mathematical operations. There are a number of other mathematical functions that are included in Python. These are shown in table 3.7.

There are also common constants built into Python:

```
e = 2.7182818 . . .
pi = 3.1415926 . . .
```

SUMMARY

The purpose of this chapter is to help the reader reach a point that he or she can write basic Python scripts. More advanced programming concepts will be introduced as they are needed throughout the book. You should not consider moving past this chapter unless you can actually make the script examples work. For those who wish to delve deeper into Python, the following sources may be of help:

www.learnpython.org/
Joshi, Prateek. Artificial Intelligence with Python. Packt Publishing.

Auffarth, Ben. Artificial Intelligence with Python Cookbook: Proven recipes for applying AI algorithms and deep learning techniques using TensorFlow 2.x and PyTorch 1.6 (p. 1). Packt Publishing.

EXERCISES

EXERCISE 1: INSTALL PYTHON

1. Open your web browser and navigate to the Downloads for Windows section of the official Python website. www.python.org/downloads/windows/.
2. Search for your desired version of Python. The version number won't change the installation substantially.

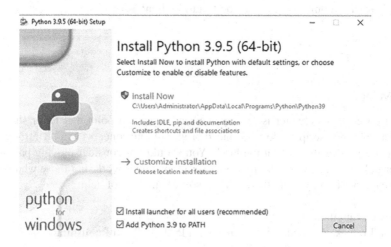

1. Open the Start menu and type "cmd." Make sure you choose to run as administrator.

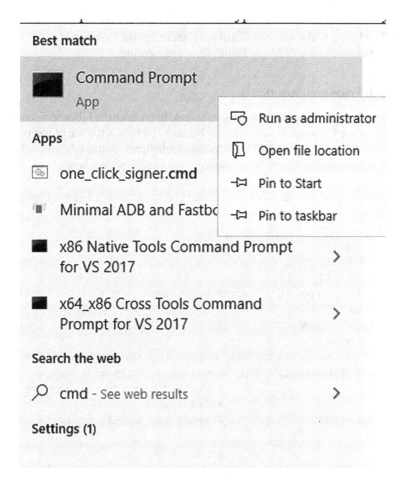

2. Select the Command Prompt application.
3. Enter pip -V in the console. If Pip was installed successfully, you should see the following output:

```
Microsoft Windows [Version 10.0.19041.985]
(c) Microsoft Corporation. All rights reserved.

C:\WINDOWS\system32>pip -V
pip 20.2.2 from c:\program files\python38\lib\site-packages\pip (python 3.8)

C:\WINDOWS\system32>
```

EXERCISE 2: HELLO WORLD

This is absurdly simple:

Open IDLE, type this in, save it, then run it.
print("Hello World")

NOTE: Make a folder off your C drivem something like C:\scripts\.
This is because you will do stuff from the command line.

EXERCISE 3: FIBONACCI SEQUENCE

A common programming exercise is to write code to produce the Fibonacci sequence. This is often used in programming classes because it involves the use of loops, print-ing to the screen, and a few other basic coding techniques. In this lab, you will write the code to produce the Fibonacci sequence. That code is shown here:

```
# function that returns a list of the numbers of the Fibonacci
series
def Fibonacci(n): # return Fibonacci series up to n
    result = []
    a, b = 0, 1
    while b < n:
        result.append(b) #append the next number to the result
        array
        a, b = b, a+b
    return result
#This calls the function and prints out the Fibonacci series
fibseries = Fibonacci(100)      # call the function. You can
                                # change the number to see different
                                # results
print (fibseries)                # print the result to the screen
```

4 More with Python

INTRODUCTION

In chapter 3, we explored the basics of Python. Carefully studying that chapter, as well as completing the exercises, should provide you with a general working knowledge of basic Python programming. In this chapter, the goal is to build upon that knowledge. There are many Python tasks you will need to understand in order to complete the machine learning Python scripts shown later in this book.

FILE MANAGEMENT

Files were briefly introduced in chapter 3. In this section we will explore file operations in more depth. It is a common task to need to read data from a file in order to perform machine learning tasks, so this is a rather critical process.

Prepare the file for reading:

In order to read in a file, you will need to create a variable for that file, and set it equal to the file you open. You will also need to determine if the file is read-only, writable,

Format:1

```
<file variable> = open(<file name>, "r")
```

The following is an example of reading a file that is in the same directory as the Python script:

```
inputFile = open("data.txt", "r")
```

If the file is in another directory then the following code will work:

```
inputFile = open("c:\\myfiles\\data.txt", "r")
```

Once you have opened the file, you have the choice to read the entire file or just individual lines of the file. Using the print function to display the file contents on the screen is an effective way to test this.

```
inputFile = open("data.txt", "r")
print (inputFile.readline())
```

Or if you want to read the first ten characters, then this code will work:

```
inputFile = open("data.txt", "r")
print (inputFile.read(5))
```

If you wish to read in the entire file, then this code will accomplish that goal:

```
inputFile = open("data.txt", "r")
```

DOI: 10.1201/9781003230588-5

```
print (inputFile.read())
```

In many cases, you might wish to read a file in, one line at a time. A loop can be an effective way of accomplishing this:

```
inputFile = open("data.txt", "r")
for i in inputFile

    print (inputFile.read())
```

Although a file is automatically closed when your program ends, it is still a good style to explicitly close your file as soon as the program is done with it.

```
inputFile.close()
```

You may also wish to prompt the user for the name of the file:
```
inputFileName = input("Enter name of input file: ")
inputFile = open(inputFileName, "r")
print("Opening file", inputFileName, "for reading.")
for i in inputFile

    print (inputFile.read())
```

```
inputFile.close()
print("Completed reading of file", inputFileName)
```

You will also find that you frequently need to read in files of a particular type. Often, data for training sets in machine learning are provided as comma-delimited files, or. csv files. Fortunately, there are several methods to easily read in such files in Python. The following is a simple script to read data from a CSV file.

```
import csv
with open("neurodata.csv", 'r') as file:
  csvreader = csv.reader(file)
  for row in csvreader:
    print(row)
```

When executed, this script reads in the file shown here:

Cerebellum, pons, mesencephalon
tegmentum, tecum, hypothalamus
The output is shown in figure 4.1.

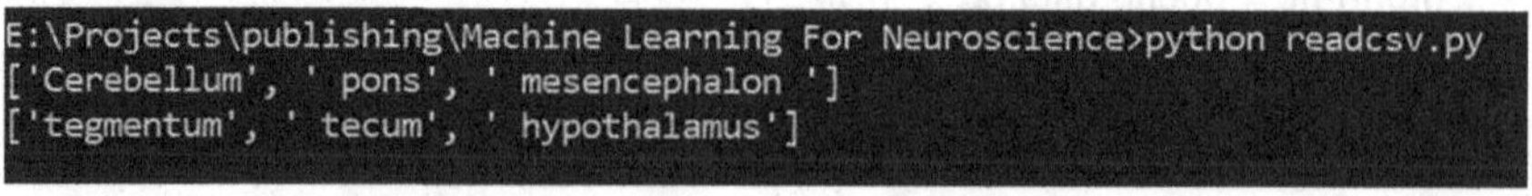

```
E:\Projects\publishing\Machine Learning For Neuroscience>python readcsv.py
['Cerebellum', ' pons', ' mesencephalon ']
['tegmentum', ' tecum', ' hypothalamus']
```

FIGURE 4.1 Reading CSV files.

```
E:\Projects\publishing\Machine Learning For Neuroscience>python writefile.py
caudal
rostral
ventral
```

FIGURE 4.2 Write to file.

You can also use the Pandas namespace, which we will discuss later in this chapter, to read in files. Here is an example:

import pandas as pd
data = pd.read_csv("neurodata.csv")

Note, our examples so far have used 'r' as the parameter. That means to open the file just for reading. In addition to the read command, you have seen there is readline, which reads a single line. There is also readlines, which reads all remaining lines in the file. You can also use a 'w' to open a file for writing. If you are working with binary data, then read is 'rb' and write is 'wb'. In our previous examples, if you wanted to write out data, you would use a script much like the one shown here.

```
import csv
somefile = open('mytextfile.txt', 'w')
anotomicaldirections = ['caudal', 'rostral', 'ventral']
for x in anotomicaldirections:
  print(x)
  somefile.write(x)
```

If the file mytextfile.txt does not exist, it will be created. The output from this script is shown in figure 4.2.

Ensure you are comfortable with reading in data from files. This will be a common task you will be repeating throughout this book and throughout your time working with machine learning.

EXCEPTION HANDLING

Errors will occur. For example, in our preceding section on file operations, if you attempt to read in a file that does not exist, an error will be generated. The typical way to address this, in most programming languages, is a try-catch exception block. In Python, rather than call it a catch block, it is called an except block. Essentially, the code you wish to execute is in the try block. The except block handles exceptions that might occur.

The following is an example of a try-catch exception block. This example uses the previous script that read in a CSV file. The difference is that in this case the file being referenced does not exist. Without exception handling, this would lead to the Python script simply crashing. Instead, the exception is trapped.

```
E:\Projects\publishing\Machine Learning For Neuroscience>python exceptionexample.py
An exception occurred
```

FIGURE 4.3 Exception handling.

```
import csv
try:
with open("missingfile.csv", 'r') as file:
csvreader = csv.reader(file)
for row in csvreader:
print(row)
except:
print("An exception occurred")
```

Due to the exception handling, the user of the script will simply see a user-friendly message, as shown in figure 4.3.

When Python encounters a try statement, it attempts to execute the statements inside the body. If there is no error, control passes to the next statement after the try . . . except. If there is an error, then the code in the except block is executed.

REGULAR EXPRESSIONS

Regular expressions are a powerful string-manipulation tool. Regular expressions are, essentially, just patterns used to match character combinations in strings. However, the flexibility in those patterns makes regular expressions a powerful tool. All modern programming languages have similar library packages for regular expressions. You can use regular expressions to search a string, replace part of a string, divide a string into sub-parts, and more.

To use regular expressions in Python you will need:

```
import re
```

The two basic functions are re.search and re.match. Search looks for a pattern anywhere in a string, whereas match looks for a match staring at the beginning. Both return none (logical false) if the pattern isn't found and a "match object" instance if it is. Here is a basic example of a regular expression.

```
import re
#this will find if the text starts with Glial and ends with
system
text = "Glial cells make up white matter in the nervous system"
x = re.search("^Glial.*system$", text)
print(x)
```

Before we look at more complex examples of regular expressions, it is first useful to understand the symbols used in regular expressions. These are shown in table 4.1.

TABLE 4.1
Regular-Expression Symbols

Character	Description	Example
[]	A set of characters	"[a-m]"
\	Signals a special sequence (can also be used to escape special characters)	"\d"
.	Any character (except newline character)	"ne.o"
^	Starts with	"^neuro"
$	Ends with	"glial$"
*	Zero or more occurrences	"ne.*o"
+	One or more occurrences	"ne.+o"
?	Zero or one occurrences	"ne.?o"
{}	Exactly the specified number of occurrences	"ne.{2}o"
|	Either or	"falls|stays

TABLE 4.2
Regular-Expression Special Characters

Character	Description	Example
\A	Returns a match if the specified characters are at the beginning of the string	"\AThe"
\b	Returns a match where the specified characters are at the beginning or at the end of a word (the "r" in the beginning is making sure that the string is being treated as a "raw string")	r"\brain" r"rain\b"
\B	Returns a match where the specified characters are present, but not at the beginning or end of a word (the "r" in the beginning is making sure that the string is being treated as a "raw string")	r"\brain" r"rain\B"
\s	Returns a match where the string contains a white space character	"\s"
\S	Returns a match where the string does not contain a white space character	"\S"
\d	Returns a match where the string contains digits (numbers from 0–9)	"\d"
\D	Returns a match where the string does not contain digits	"\D"
\w	Returns a match where the string contains any word characters (letters, digits, and the underscore _ character)	"\w"
\W	Returns a match where the string does not contain any word characters	"\W"
\Z	Returns a match if the specified characters are at the end of the string	"neuron\Z"

There are also special characters that can be used in regular expressions. These are shown in table 4.2.

The following is a more robust version of the earlier script.

```
import re
#this will find if the text starts with Glial and ends with make
text = "Glial cells make up white matter in the nervous system."
```

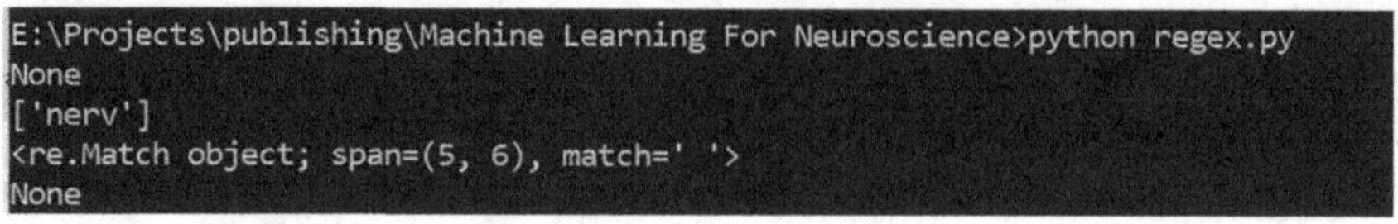

```
E:\Projects\publishing\Machine Learning For Neuroscience>python regex.py
None
['nerv']
<re.Match object; span=(5, 6), match=' '>
None
```

FIGURE 4.4   Regex.

```
x = re.search("^Glial.*make$", text)
print(x)
#find all instances of 'nerv
x = re.findall("nerv", text)
print(x)
#find the first whitespace
x = re.search("\s", text)
print(x)
x = re.search('d', text)
print(x)
```

The output from this script is shown in figure 4.4.

Familiarizing yourself with the essentials of this section will give you the basics of regular expressions. As you may already surmise, one could certainly go much further with regular expressions. For those readers who wish to dive deeper into this topic, I suggest the following resources:

- https://docs.python.org/3/howto/regex.html
- https://developers.google.com/edu/python/regular-expressions
- https://pynative.com/python/regex/

INTERNET PROGRAMMING

You may find that from time to time you need to access internet resources. This may be to locate and obtain a dataset, to upload results, or related purposes. For this reason, a basic understanding of internet programming is provided in this section.

The first item to familiarize yourself with is Urllib. It has several objects within it for various internet activities. There is Urllib.request for opening URLs and reading the content. There is Urllib.parse for parsing URL content. Then the Urllib.error is for exceptions specific to Urllib. Finally, there is Urllib.robotparser for reading and parsing robot.txt files. The following example should help clarify Urllib.

```
import urllib.request
#this website returns your public IP address
external _ ip = urllib.request.urlopen('https://ident.me').read().
decode('utf8')
# printout that IP
print(external _ ip)
```

When executed, this will print to the screen your public IP address. A simple search of the internet for Python scripts to print your public IP address will reveal many other methods to accomplish this goal. Our purpose, however, is to introduce you to the Urllib library. If for some reason, this code does not execute, it is most likely due to not having Urllib installed on your machine. You can correct that by installing it:

pip install urllib
The following code provides a bit more utilization of Urllib and combines it with the regular expressions discussed in the previous section.

```
from urllib.request import urlopen
import re
#connect to a URL
website = urlopen("www.chuckeasttom.com")
#read html code. The decode statement takes the
#byte input and converts it to string
html = website.read().decode('utf8')
#use re.findall to get all the links
links = re.findall('"((http|ftp)s?://.*?)"', html)
print (links)
```

I used my own website as an example, but you can use any URL you wish and extract links from that website.

INSTALLING MODULES

One of the strengths of Python is the plethora of packages one can add. There are packages for everything. You will use the Package Installer for Python, or PIP. It is usually quite simple to install a package. Sometimes you might need to first upgrade your installer. That, too, is rather easy.

pip install—upgrade pip
This is shown in figure 4.5.

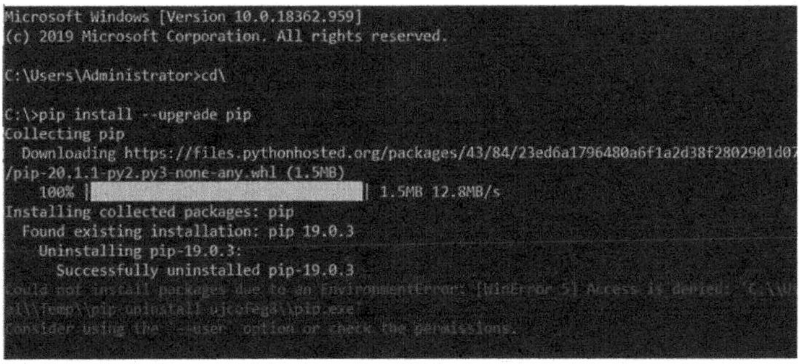

FIGURE 4.5   Upgrade Pip.

Pip is the preferred installer program. Starting with Python 3.4, it is included by default with the Python binary installers.

```
python -m pip install SomePackage
python -m pip install SomePackage==1.0.4 # specific version
python -m pip install "SomePackage>=1.0.4" # minimum version
python -m pip install-upgrade SomePackage
```

SPECIFIC MODULES

Operating System Module

The operating system module is designed to give you access to the underlying operating system your script is executing on. This allows you to access files, directories, and perform related activities.

Generic operating system interface

```
getcwd()      Get the current directory name
listdir()     List the files in a directory
chown()       Change the ownership of a file
chmod()       Change the permissions of a file
rename()      Rename a file
remove()      Delete a file
mkdir()       Create a new directory
system()      Execute command in a subshell
```

NumPy

The NumPy packet is specifically for mathematical operations. It is powerful and flexible. As you can probably surmise, it is frequently used in machine learning tasks.

One of the items in NumPy that is frequently used in machine learning is the array, called ndarray. This is a multidimensional array. The properties and methods are shown in table 4.3.

The following is an example of using the ndarray datatype.

```
import numpy as np
a = np.arange(15).reshape(3, 5)
a
array([[0, 1, 2, 3, 4],
[5, 6, 7, 8, 9],
[10, 11, 12, 13, 14]])
a.shape
(3, 5)
a.ndim
2
a.dtype.name
'int64'
a.itemsize
```

```
8
a.size
15
```

Structured datatypes can be created using the function numpy.dtype. There are four forms of this specification, they are:

1. A string of comma-delimited dtypes
2. A list of tuples with one tuple per field. Each individual tuple is of the form (fieldname, datatype, shape), with shape being a tuple of integers that describe the subarray shape. The shape is optional. The fieldname is a string, and the datatype is obviously the datatype. Something like:
   ```
   np.dtype([('a', 'np.float32', (3, 3))])
   ```
3. A dictionary of field parameter arrays
4. A dictionary of field names

You can also read in from text files using numpy, though this is a less common application. It looks something like this:

```
import numpy as np
np.genfromtxt("mycsvfile.txt", delimiter=",", usemask=True)
Keras
```

Keras is often used for deep learning. You will use it with many of the examples for neuroscience machine learning later in this book. Keras is an open-source library used to provide artificial neural networks to Python. There are built-in objects and methods for all the neural network elements you will learn in chapters 9, 10, and 11.

The Keras functional API is a way to create models. The functional API can handle models with nonlinear topology, models with shared layers, and models with

---

TABLE 4.3
Numpy Properties and Methods

Property/Method	Usage
ndarray.ndim	The number of axes (dimensions) of the array.
ndarray.size	The total number of elements of the array. This is equal to the product of the elements of shape.
ndarray.shape	The dimensions of the array. This is a tuple of integers indicating the size of the array in each dimension. For a matrix with n rows and m columns, shape will be (n, m). The length of the shape tuple is therefore the number of axes, ndim.
ndarray.dtype	An object describing the type of the elements in the array. One can create or specify dtypes using standard Python types.
ndarray.data	The buffer containing the actual elements of the array. Normally, we won't need to use this attribute because we will access the elements in an array using indexing facilities.

multiple inputs or outputs. Keras is a deep learning API written in Python, running on top of the machine learning platform TensorFlow. It was developed with a focus on enabling fast experimentation.

tf.keras offers a higher API level, with three different programming models: Sequential API, Functional API, and Model Subclassing.

A model has a life cycle, and this very simple knowledge provides the backbone for both modeling a dataset and understanding the tf.keras API.

The five steps in the life cycle are as follows:

Define the model.
Compile the model.
Fit the model.
Evaluate the model.
Make predictions.

Pandas

Pandas is a library for data analysis. You saw it earlier in this chapter used to read in a CSV file. Here is a sample of Pandas with a dataset:

```
import pandas
mydataset = {
'icecream': ["Chocolate", "Vanilla", "Strawberry"],
'scoops': [1, 2, 3]
}
myvar = pandas.DataFrame(mydataset)
print(myvar)
```

The entire dataset can be treated as a single object. This is quite convenient when using data as a training set for machine learning. Pandas provides methods to handle data within your Python scripts, but also ways to import data. The previous example uses a dataset. You can also use Pandas to work with a series. A Pandas series is very much like a column in a table. Recall the review of linear algebra in Chapter 1. Pandas' series are a good way to encode vectors. Here is a simple example.

```
import pandas as pd
a = [1, 2, 3]
myvar = pd.Series(a)
print(myvar)
```

If you don't specify a label for the elements in the series, then the elements are labeled with their index number. This index is a zero-based array. So, if you wish to print the second item in the series you would use:

```
print(myvar[1])
```

You can choose to label the elements in the series if you wish. The following sample script assigns labels, albeit not particularly creative ones.

```
import pandas as pd
a = [1, 2, 3]
myvar = pd.Series(a, index = ["first", "second", "third"])
print(myvar)
```

You saw, earlier in this chapter, Pandas being used to read in a CSV file; it also works with JSON (JavaScript Object Notation) files. The following is a simple example of Pandas reading in a JSON file.

```
import pandas as pd
df = pd.read _ json('data.json')
print(df.to _ string())
```

In the preceding example script, you can also choose to print only parts of the data read in. For example, if you wish to print the first five rows, then:

```
import pandas as pd
df = pd.read _ json('data.json')
print(df.head())
```

Or perhaps you wish to print the last five rows. Then you can use this script:

```
import pandas as pd
df = pd.read _ json('data.json')
print(df.tail())
```

One of the aspects of Pandas that you will find very useful in machine learning is the ability to clean up data you import. The dataset you acquire may not be in the format you need. Consider a. csv file that you need to have in date/time format. The following script will change the format to date/time.

```
import pandas as pd
df = pd.read _ csv('data.csv')
df['Date'] = pd.to _ datetime(df['Date'])
print(df.to _ string())
```

Perhaps your dataset has null values. These won't be particularly useful in machine learning, so you may wish to drop them. If, for example, some of the date/time fields are null, you can drop those with one small addition to our script.

```
import pandas as pd
df = pd.read _ csv('data.csv')
df['Date'] = pd.to _ datetime(df['Date'])
df.dropna(subset=['Date'], inplace = True)
print(df.to _ string())
```

As you move forward in this book, particularly in chapters 9 through 11, you will find many of the code examples use Pandas. Ultimately, machine learning is all about data. Whether it is training data, real-world data that is being evaluated, or output from a machine learning algorithm, it is all about data.

Scikit-Learn

This library is specifically for machine learning. It has built into it many of the algorithms we will discuss in chapters 9 through 11, such as k-means clustering, gradient boosting, and random forests. The following is a common example of using scikit-learn.

```
# load the iris dataset as an example. The Iris dataset consists
of 3 different types of
# flower irises including information about sepal and petal. It
is often used when learning

# machine learning.
from sklearn.datasets import load _ iris #Note you can simply
ask SciKit Learn to import
#this dataset. It will download it from its well
#know data source on the web.
iris = load _ iris()
# store the feature matrix (X) and response vector (y)
X = iris.data
y = iris.target
# store the feature and target names
feature _ names = iris.feature _ names
target _ names = iris.target _ names
# printing features and target names of our dataset
print("Feature names:", feature _ names)
print("Target names:", target _ names)
# X and y are numpy arrays
print("\nType of X is:", type(X))
# printing first 5 input rows
print("\nFirst 5 rows of X:\n", X[:5])
```

PyTorch

The PyTorch library is widely used in machine learning. It is particularly used in natural language processing and computer vision. PyTorch was originally developed by Facebook (now Meta). PyTorch defines a class named Tensor that is used to work with multidimensional arrays of numbers. This is similar to the arrays in NumPy, but the PyTorch arrays work well with other platforms, including several graphics processing units (GPUs).

PyTorch has a number of useful components in it. For example, the Optim module has built-in optimization algorithms that are very useful in building artificial neural networks. The Distributed module supports three different backends (Gloo, MPI, and NCCL). The Gloo backend is a communications library that includes several algorithms that can be used in machine learning. We will see this algorithm later in this book. The MPI library is meant for use in high-performance computing (HPC) and includes support for multi-threaded applications and other features useful in machine learning. Finally, the NCCL backend (often this is pronounced "nickel") is a library of routines used to communicate with GPUs.

Torch.NN is a module you will see a lot in this book. As the name suggests, this module supports neural networks. There are objects for activation functions, layers, distance functions, and more. Torch.FFT contains support for fast Fourier transforms. These are widely used in machine learning. If you are not familiar with FFT or Fourier transforms in general, a brief explanation is provided. Fourier transforms convert a signal from its original form to a representation in a frequency domain. The reverse can also be accomplished with Fourier transforms. For those readers who would like a more rigorous mathematical definition, a Fourier transform can be defined as a technique for mapping a function from its original function space into a different function space.

There are a number of variations of Fourier transforms, including discrete Fourier transforms (DFT) and fast Fourier transforms (FFT). The FFT was originally applied to audio signals, and it converts the signal into individual spectral components. FFTs and other Fourier transforms are often useful in neural networks, as you will see later in this book.

WMI

If you are running your scripts on a Windows computer, then the WMI module will be useful to you. WMI, or Windows Management Interface, was developed by Microsoft to facilitate interacting with the Windows operating system. You can do a wide range of things with this module. First you need to install the WMI library, as shown in figure 4.6.

We will first examine a simple WMI Python script that retrieves information about all running processes on your Windows computer. The script is shown in figure 4.7:

```
E:\Projects\publishing\Machine Learning For Neuroscience>pip install wmi
Requirement already satisfied: wmi in c:\python310\lib\site-packages (1.5.1)
Requirement already satisfied: pywin32 in c:\python310\lib\site-packages (from wmi) (304)

E:\Projects\publishing\Machine Learning For Neuroscience>
```

FIGURE 4.6   Install WMI.

wmione.py - E:/Projects/publishing/Machine Learning For Neuroscience/wmione.py (3.10.4)   —

File   Edit   Format   Run   Options   Window   Help

```
import wmi
myWMI = wmi.WMI ()

for process in myWMI.Win32_Process ():
  print (process.ProcessId, process.Name)
```

FIGURE 4.7   WMI script.

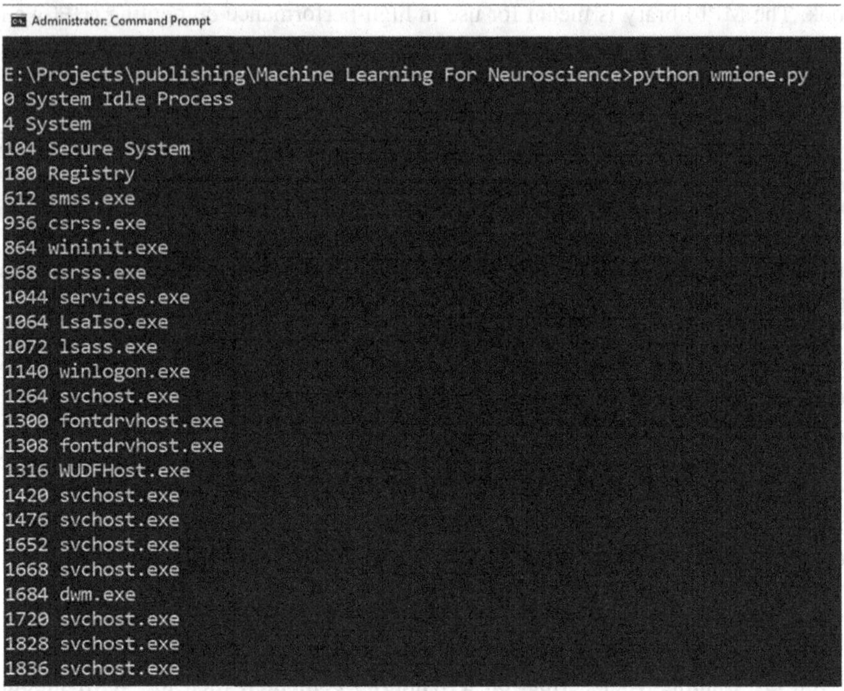

FIGURE 4.8 Executing WMI script.

As you can see, this is a very simple script. When you execute the script, it may take a few moments for it to retrieve all current running processes. The result is shown in figure 4.8.

The following code shows the general features of WMI, or at least those that are most commonly used.

```
import wmi
c = wmi.WMI()
for os in c.Win32_OperatingSystem():
print (os.Caption)
for service in c.Win32_Service(Name="seclogon"):
result, = service.StopService()
if result == 0:
print ("Service", service.Name, "stopped")
else:
print ("Some problem")
break
else:
print ("Service not found")
for disk in c.Win32_LogicalDisk(DriveType=3):
print (disk)
```

```
wql = "SELECT Caption, Description FROM Win32 _ LogicalDisk WHERE
DriveType <> 3"
for disk in c.query(wql):
print (disk)
for method _ name in os.methods:
method = getattr(os, method _ name)
print (method)
result, = c.Win32_Share.Create(Path="c:\\temp", Name="temp",
Type=0)
if result == 0:
print ("Share created successfully")
else:
print ("Problem creating share: %d" % result)
```

When you execute this code you will see results much like what is shown in figure 4.9.

Working with these two basic scripts should give you a general idea of what you can do with WMI. There is much more that can be found at the Python WMI website.[1]

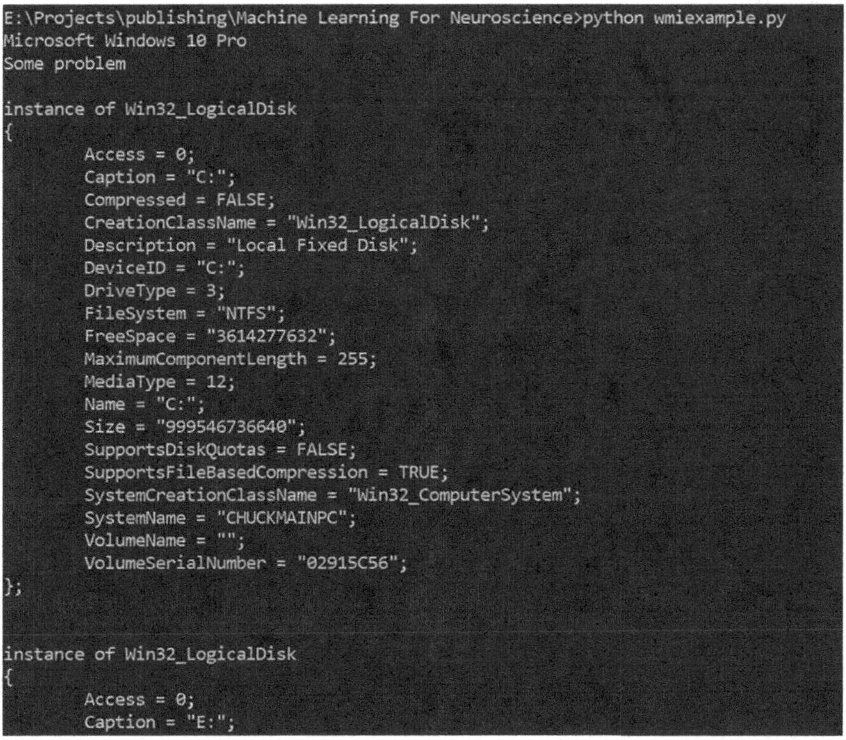

```
E:\Projects\publishing\Machine Learning For Neuroscience>python wmiexample.py
Microsoft Windows 10 Pro
Some problem

instance of Win32_LogicalDisk
{
        Access = 0;
        Caption = "C:";
        Compressed = FALSE;
        CreationClassName = "Win32_LogicalDisk";
        Description = "Local Fixed Disk";
        DeviceID = "C:";
        DriveType = 3;
        FileSystem = "NTFS";
        FreeSpace = "3614277632";
        MaximumComponentLength = 255;
        MediaType = 12;
        Name = "C:";
        Size = "999546736640";
        SupportsDiskQuotas = FALSE;
        SupportsFileBasedCompression = TRUE;
        SystemCreationClassName = "Win32_ComputerSystem";
        SystemName = "CHUCKMAINPC";
        VolumeName = "";
        VolumeSerialNumber = "02915C56";
};

instance of Win32_LogicalDisk
{
        Access = 0;
        Caption = "E:";
```

FIGURE 4.9 Second WMI script.

PIL

The Python Imaging Library is often used when working with images. In later chapters you will frequently be working with medical imaging data from brain scans (MRI, EEG, etc.). An introduction to the basics of image handling in PIL will make those coding examples go easier.

> One of the most common tasks is simply opening an image file. This is done with
> PIL.Image.open(fp, mode = 'r', formats=None)
> The fp is a string for the filename. The second parameter is mode which should be 'r'. The final parameter is for file format. None denotes trying all file formats.
> You can also open an image using the Image object in PIL as shown here:
> with Image.open("brainphoto.png") as im
> To open an image and rotate it 45 degrees, the following code is used.

```
from PIL import Image
with Image.open("brainphoto.png") as im:
im.rotate(45).show()
```

When executing that code, you will see an image such as what is shown in figure 4.10.

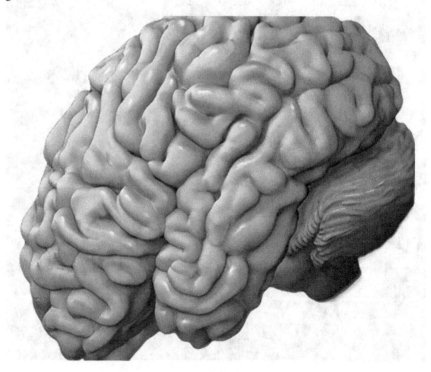

FIGURE 4.10   Load image with PIL.

Note there is much more you can do with the Image class; for more details consult the Pillow documentation.[2]

You can also create images using code such as:

```
PIL.Image.new(mode, size, color=0)
```

Mode is the mode for your new image. Common modes are listed here:

L: 8 bit pixels in black and white
RGB: 3 x 8 pixels color
CMYK: 4 x 8 pixels color
YCbCr: 3 x 8 pixels color video
Size is height and width; and color is obvious, black is the default.
You can even generate the Mandelbrot set using:
PIL.Image.effect_mandelbrot(size, extent, quality)
Size is again height and width, extent is the extent to cover and is given as a four-tuple, and quality is obviously the quality of the image. The following code will display the Mandlebrot set:

```
from PIL import Image
size = (512, 512)
extent = (-3, -2.5, 2, 2.5)
quality = 100
im = Image.effect_ mandelbrot(size, extent, quality)
im.show()
```

When executing that code, you will see an image such as what is shown in figure 4.11.

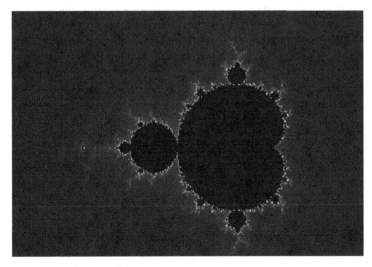

FIGURE 4.11   Mandelbrot set from PIL.

The Python Image Library is replete with capabilities to create images, or to read in images. This allows you to have a very wide range of image functionality.

MATPLOTLIB

This is a library that you will see used in many of the machine learning examples later in this book. Matplotlib is a library that allows you to create all sorts of images, both static and animated. You can also export your graphs to a wide range of formats. Let us begin by examining the simple script shown here:

```python
# You need the matplotlib to create graphs
import matplotlib.pyplot as plt
# x axis values
x = [1,2,3]
# corresponding y axis values
y = [6,5,4]
# plotting the points
plt.plot(x, y)
# naming the two axis
plt.xlabel('X axis')
plt.ylabel('Yaxis')
# Title the graph
plt.title('Test Graph')
# function to show the plot
plt.show()
```

When you execute this script, you will see a graph, much like what is shown in figure 4.12.

Notice that the toolbar at the bottom is created for you automatically. If you click on the save icon you will be given several choices; those are shown in figure 4.13.

It is also quite easy to create other sorts of charts, such as a bar chart. The following code shows a simple bar chart using Matplotlib.

```python
import matplotlib.pyplot as plt
# x-coordinates of left sides of bars
left = [1, 2, 3, 4, 5]
# heights of bars
height = [10, 15, 22, 31, 1]
# labels for bars
tick_label = ['one', 'two', 'three', 'four', 'five']
# plot a bar chart
plt.bar(left, height, tick_label = tick_label,
width = 0.8, color = ['red', 'blue'])
# Label the axis
plt.xlabel('X axis')
plt.ylabel('Y axis')
```

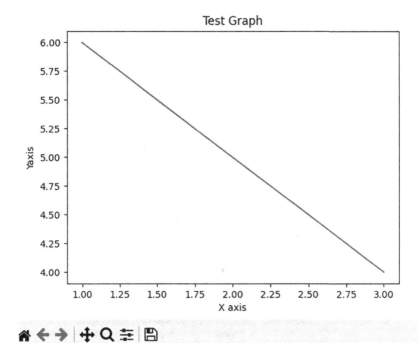

FIGURE 4.12 Basic graph from Matplotlib.

Figure_1.png

Portable Network Graphics (*.png)
Portable Network Graphics (*.png)
Encapsulated Postscript (*.eps)
Joint Photographic Experts Group (*.jpeg;*.jpg)
Portable Document Format (*.pdf)
PGF code for LaTeX (*.pgf)
Postscript (*.ps)
Raw RGBA bitmap (*.raw;*.rgba)
Scalable Vector Graphics (*.svg;*.svgz)
Tagged Image File Format (*.tif;*.tiff)
WebP Image Format (*.webp)

FIGURE 4.13 Saving a graph.

```
# Title the bar chart
plt.title('Sample Bar Chart')
# function to show the plot
plt.show()
```

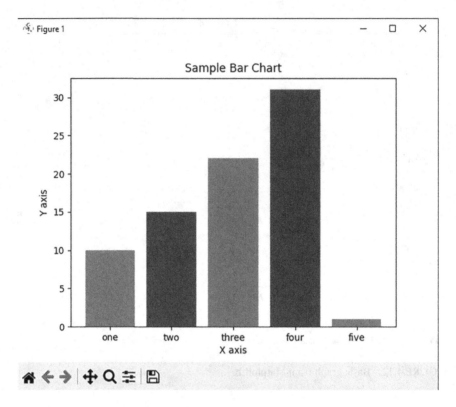

FIGURE 4.14 Matplotlib bar chart.

When you execute that code, you should see something much like what is shown in figure 4.14.

In both the code samples in this section, arbitrary values where graphed. Obviously, when you are working on machine learning for neuroscience, you will have actual data you wish to graph. The purpose of the preceding code samples was just to help you get comfortable with Matplotlib.

TENSORFLOW

You are going to work a lot with TensorFlow later in this book. Much of the use of TensorFlow will make more sense once you have read the chapters regarding machine learning. However, it is worthwhile to introduce you to TensorFlow now. TensorFlow was first developed by the GoogleBrain team for internal use. It was released to the public in late 2015. The current version is Tensorflow 2.0. TensorFlow provides Python and C APIs. It does not guarantee backwards compatibility. There are packages for C++, Java, Go, Swift, Matlab, C#, and others. However, we will focus on Python.

```
pip install tensorflow
```

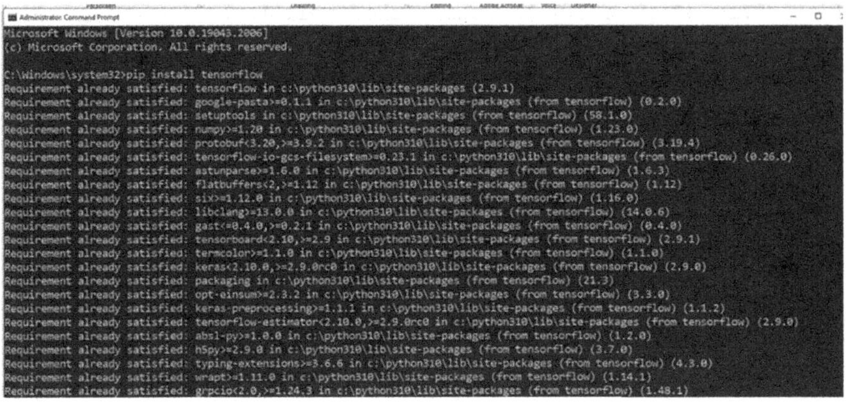

FIGURE 4.15 Install TensorFlow.

This is shown in figure 4.15.

Or if you have a GPU you wish to use:
pip install tensorflow-gpu

This will take a substantial amount of time. You may even think something has gone wrong. Just be patient. It can take 20 to 30 minutes in some cases.

TensorFlow is based on graph-based computation. A computational graph is a network of nodes and edges. This is an alternative way of conceptualizing mathematical calculations. A tensor is a mathematical entity with which to represent different properties, similar to a scalar, vector, or matrix. Tensors are multidimensional arrays that have some dynamic properties. Tensors can be of two types: constant or variable.

The API known as TensorFlow Core provides fine-grained lower-level functionality. Because of this, this low-level API offers complete control while being used on models.

High-level API: These APIs provide high-level functionalities that have been built on TensorFlow Core and are comparatively easier to learn and implement. Some high-level APIs include Estimators, Keras, TFLearn, TFSlim, and Sonnet.

The following is a basic TensorFlow Python script:

```python
#!/usr/bin/python
# this is a basic tensor flow project
import tensorflow as tf
mnist = tf.keras.datasets.mnist
# load the data set
(x_train, y_train), (x_test, y_test) = mnist.load_data()
x_train, x_test = x_train / 255.0, x_test / 255.0
#Build the tensorflow tf.keras.Sequential model by stacking layers.
model = tf.keras.models.Sequential([
tf.keras.layers.Flatten(input_shape=(28, 28)),
```

```
tf.keras.layers.Dense(128, activation='relu'),
tf.keras.layers.Dropout(0.2),
tf.keras.layers.Dense(10)
])
#For each example the model returns a vector of "logits"
#or "log-odds" scores, one for each class.
predictions = model(x_train[:1]).numpy()
predictions
#The tensorflow tf.nn.softmax function converts these logits
#to "probabilities" for each class:
tf.nn.softmax(predictions).numpy()
#The losses.SparseCategoricalCrossentropy loss takes a vector of
logits
#and a True index and returns a scalar loss for each example.
loss_fn = tf.keras.losses.SparseCategoricalCrossentropy(from_
logits=True)
loss_fn(y_train[:1], predictions).numpy()
model.compile(optimizer='adam',
loss=loss_fn,
metrics=['accuracy'])
model.fit(x_train, y_train, epochs=5)
#The Model.evaluate method checks the models performance
model.evaluate(x_test, y_test, verbose=2)
```

Rank simply indicates the number of directions required to describe the properties of an object, meaning the dimensions of the array contained in the tensor itself. A scaler is rank 0, a vector rank 1, a matrix rank 2, etc.

The **shape** of a tensor represents the number of values in each dimension. So, a three-element vector would have a shape of [3]. A 4 x 4 matrix would be shape [4,4]. New with TensorFlow 2.0 is the tf.function capability, which converts relevant Python code into a TensorFlow graph.

The devices in TensorFlow are identified with the string/device: < device_type >: < device_idx >. In the last output, CPU and GPU denote the device type, and 0 denotes the device index.

Importing a data set is easy. First you have to install it:
pip install tensorflow-datasets

This is shown in figure 4.16
Then the script is actually quite simple:

```
import tensorflow as tf
import tensorflow_datasets as tfds
# See all registered datasets
builders = tfds.list_builders()
print(builders)
```

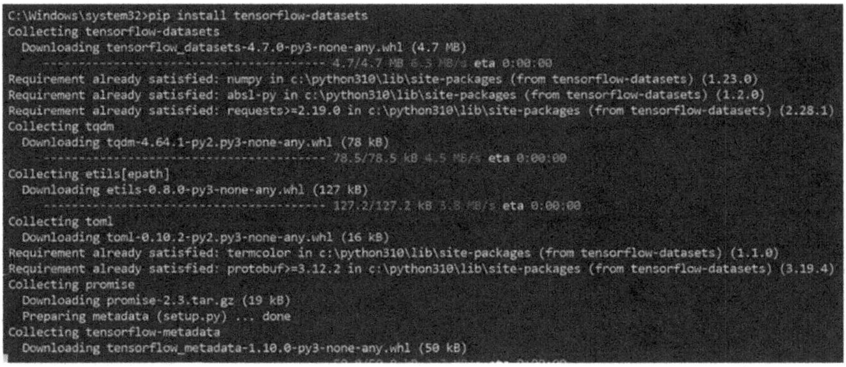

FIGURE 4.16 Tensorflow datasets.

THE ZEN OF PYTHON

As you continue through this book, you will get more experience writing Python. First with basic Python scripts, and then later with scripts specific for machine learning applied to neuroscience. However, it is also hoped that your Python programming will improve as you move forward. There is a "Zen of Python," which enumerates the best way to do Python programming:

- Beautiful is better than ugly
- Explicit is better than implicit
- Simple is better than complex
- Complex is better than complicated
- Readability counts

All of this comes down to an even simpler concept. Write your code so any competent Python programmer can read it. Even a novice. Your algorithms should never be any more complex than they need to be.

ADVANCED TOPICS

This section covers topics that are usually a part of undergraduate computer science curriculum, but are often missed by programmers who did not have a formal computer science education. It is certainly possible to write Python scripts without this knowledge. It is also possible to perform machine learning tasks without this knowledge. However, if you wish to delve deeper into machine learning algorithms, at some point, you will need this fundamental knowledge.

DATA STRUCTURES

A data structure is a formal way of storing data that also defines how the data is to be processed. This means that the definition of any given data structure must detail

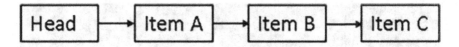

FIGURE 4.17   A list.

how the data is going to be stored, and what methods are available for moving data into and out of that particular data structure. There are a number of well-defined data structures that are used in various situations. Different data structures are suited for different applications. In this section, we will examine several of the more commonly known data structures. Data structures are fundamental to computer science.

Lists

A list is one of the simplest of data structures. It is a structure consisting of an ordered set of elements, each of which may be a number, another list, etc. A list is usually denoted something like you see here:

(a, b, c, d, . . .)

Many other data structures are some sort of extension of this basic concept. A list can be either homogenous or heterogeneous. That simply means that the elements in the list can all be the same type or not. If you have a background in set theory, then this should look very familiar to you. A list is just a set. Figure 4.17 shows the structure of a list.

The most common implementation of a list, in most programming languages, is the array. And in most programming languages, an array is usually homogenous (i.e., this means all elements of the array are of the same data type). It should be noted that several object-oriented programming languages offer a type of list known as a collection, that is heterogeneous (i.e., elements of that list may be of diverse data types).

Queue

A queue is simply a special case of a list. It stores data in the same way that a list does; it is also often implemented as simply an array. The difference between a list and a queue is in the processing of the data contained in either data structure. A queue has a more formal process for adding and removing items than the list does. In a queue, data is processed on a first-in-first-out basis (FIFO). Often there is some numerical pointer designating where the last data was input (often called a tail) and where the last data was extracted (often called a head). Putting data into the queue is referred to as enqueueing, and removing it is dequeuing. Figure 4.18 shows the structure of a queue.

Queues are a very common data structure. If you consider for a moment a printer, you will note that a printer processes incoming print commands using a queue.

Stack

The stack is a data structure that is a special case of the list. With the stack, elements may be added to or removed from the top only. Adding an item to a stack is referred

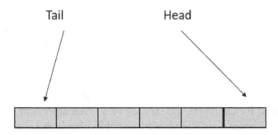

FIGURE 4.18   A queue.

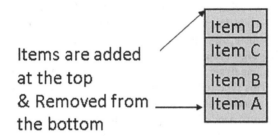

FIGURE 4.19   FIFO structure.

to as a push, and removing an item is referred to as a pop. In this scenario, the last item added must be the first one removed, or last-in-first-out (LIFO). A good analogy is to consider a stack of plates. Because the last item in is the first out, this data structure does not require a pointer. It is probably worthwhile to compare LIFO with FIFO (first-in-first-out). The FIFO data structure requires that the first item put in will also be the first one out. This is similar to the stack, but the popping and pushing take place on different ends. This is shown in figure 4.19

Linked List

A linked list is a data structure wherein each item has a pointer (or link) to the next item in the list, or the item preceding it, but not both (a doubly linked list, which will be discussed later, does both). This can be quite useful since if an item is displaced, for any reason; it is aware of the next item in the list. Each item knows what comes after it, or what comes before it, but not both.

A doubly linked list is a data structure wherein each item has a pointer (or link) to the item in front of it, and the item behind it. This data structure is the logical next step after the linked list. It is highly efficient in that any data point has links to the preceding and succeeding data points.

ALGORITHMS

Much of this book is devoted to helping you learn specific machine learning algorithms, in order to apply those algorithms to neuroscience. It is appropriate to have an understanding of what algorithms are. An algorithm is simply a systematic way of solving a problem. A recipe for an apple pie is an algorithm. If you follow the procedure, you get the desired results. Algorithms are a routine part of computer programming and an integral part of computer science.

It is not enough to simply have some algorithm to accomplish a task. Computer science seeks to find the most efficient algorithm. Therefore, it is also important that we have a clear method for analyzing the efficacy of a given algorithm. When considering any algorithm, if the desired outcome is achieved, then clearly, the algorithm worked. But the real question is, how well did it work? If you are sorting a list of ten items, the time it takes to sort the list is not of particular concern. However, if your list has 1 million items, then the time it takes to sort the list, and hence the algorithm you choose, is of critical importance. This is especially true in machine learning. You may have an algorithm that works and achieves a high level of accuracy, but is simply not efficient enough for practical use. Fortunately, there are well-defined methods for analyzing any algorithm.

When analyzing algorithms, we often consider the asymptotic upper and lower bounds. Asymptotic analysis is a process used to measure the performance of computer algorithms. This type of performance is based on the complexity of the algorithm. Usually this is a measure of either the time it takes for an algorithm to work, or the resources (memory) needed. It should be noted that one usually can only optimize time or resources, but not both. The asymptotic upper bound is simply the worst-case scenario for the given algorithm, whereas the asymptotic lower bound is a best case.

Some analysts prefer to simply use an average case; however, knowing the best case and worst can be useful in some situations. In simple terms, both the asymptotic upper bound and lower bounds must be within the parameters of the problem you are attempting to solve. You must assume that the worst-case scenario will occur in some situations.

Perhaps the most common way to formally evaluate the efficacy of a given algorithm is Big O notation. This method is a measure of the execution of an algorithm, usually the number of iterations required, given the problem size n. In sorting, algorithms n is the number of items to be sorted. Stating some algorithm $f(n) = O(g(n))$ means it is less than some constant multiple of $g(n)$. The notation is read, "f of n is big oh of g of n." This means that saying an algorithm is 2N means it will have to execute two times the number of items on the list. Big O notation essentially measures the asymptotic upper bound of a function. Big O is also the most often used analysis.

Big O notation was first introduced by the mathematician Paul Bachmann in his 1892 book *Analytische Zahlentheorie*. The notation was popularized in the work of another mathematician named Edmund Landau. Because Landau was responsible for popularizing this notation, it is sometimes referred to as a Landau symbol.

Omega notation (Ω) is the opposite of Big O notation. It is the asymptotic lower bound of an algorithm and gives the best-case scenario for that algorithm. It gives you the minimum running time for an algorithm. Theta notation (Θ) combines Big O and Omega to give the average case (average being the arithmetic mean in this situation) for the algorithm. In our analysis, we will focus heavily on the Theta, also often referred to as the Big O running time. This average time gives a more realistic picture of how an algorithm executes.

SUMMARY

This chapter builds on the information in chapter 3. The goal is to provide the reader with a strong working knowledge of Python. This chapter introduced you to file management and exception handling, which will both be critical for machine learning projects you create later in this book. Regular expressions were also introduced. These are essential to almost any type of programming, including machine learning. Then additional topics, such as internet programming and installing modules, were covered. Combined with chapter 3, the material in this chapter should provide you a working knowledge of Python programming.

EXERCISES

EXERCISE 1: INSTALL TENSORFLOW

Assumes you already have Python installed. If not:
www.python.org/downloads/
pip install—upgrade pip

```
Microsoft Windows [Version 10.0.18362.959]
(c) 2019 Microsoft Corporation. All rights reserved.

C:\Users\Administrator>cd\

C:\>pip install --upgrade pip
Collecting pip
  Downloading https://files.pythonhosted.org/packages/43/84/23ed6a1796480a6f1a2d38f2802901d07
/pip-20.1.1-py2.py3-none-any.whl (1.5MB)
    100% |                                    | 1.5MB 12.8MB/s
Installing collected packages: pip
  Found existing installation: pip 19.0.3
    Uninstalling pip-19.0.3:
      Successfully uninstalled pip-19.0.3
Could not install packages due to an EnvironmentError: [WinError 5] Access is denied: 'C:\\Us
al\\Temp\\pip-uninstall-ujcofeg8\\pip.exe'
Consider using the  --user' option or check the permissions.
```

Common error: Could not install packages due to an EnvironmentError: [WinError 5] Access is denied: 'C:\\Users\\ADMINI~1\\AppData\\Local\\ Temp\\pip-uninstall-ujcofeg8\\pip.exe'
Consider using the '--user' option or check the permissions
If you get that error correct with
pip install --upgrade pip --user

```
C:\>pip install --upgrade pip --user
Requirement already up-to-date: pip in c:\python\lib\site-packages (20.1.1)

C:\>
```

pip install tensorflow

This will take a VERY long time. You will think something has gone wrong. Just be patient. It can take 20 to 30 minutes in some cases

```
C:\>pip install tensorflow
Collecting tensorflow
  Downloading tensorflow-2.2.0-cp37-cp37m-win_amd64.whl (459.2 MB)
                            | 152.4 MB 6.4 MB/s eta 0:00:48
```

EXERCISE 2: REGULAR EXPRESSIONS AND EXCEPTION HANDLING

First, simply code the regular-expression example show earlier in this chapter. Make sure it functions as written. That code is shown here again.

```
import re
#this will find if the text starts with Glial and ends with make
text = "Glial cells make up white matter in the nervous system."
x = re.search("^Glial.*make$", text)
print(x)
#find all instances of 'nerv'
x = re.findall("nerv", text)
print(x)
#find the first whitespace
x = re.search("\s", text)
print(x)
x = re.search('d', text)
print(x)
```

Once you can make this work as shown, you will need to modify it:

1. First try a few different examples of using regular expressions. The goal is to get comfortable with using regular expressions.
2. Now add exception handling to the script.
3. (Optional): For those readers seeking a more substantive challenge, modify the script so that it reads in a text file and uses regular expressions to search that text-file data.

EXERCISE 3

Use the following code to create a scatterplot.

```
import matplotlib.pyplot as plt
# x axis values
x = [1,2,3,4,5,6,7,8,9,10]
# y axis values
```

```
y = [4,4,5,7,6,8,5,10,7,6]
# plotting points as a scatter plot
plt.scatter(x, y, label= "Asterisks", color= "blue",
            marker= "*", s=20)
# Label the axis
plt.xlabel('X axis')
plt.ylabel('Y axis')
# plot title
plt.title('Example Scatter Plot')
# function to show the plot and the legend
plt.legend()
plt.show()
```

NOTES

1. https://pypi.org/project/WMI/
2. https://pillow.readthedocs.io/en/stable/reference/Image.html

Section II

Required Neuroscience

5 General Neuroanatomy and Physiology

INTRODUCTION

In order to work in machine learning as applied to neuroscience, one obviously requires a knowledge of neuroscience. This begins with understanding the general neuroanatomy and physiology. The approach in this chapter will be to start with a broad overview of major anatomical structures and gradually add more detail as the chapter progresses. Certainly, some textbooks begin with cellular operations and build up to larger anatomical structures. And one can make an argument for that approach. However, my own experience indicates that it is easier for novices to begin with large structures that can be related directly to functions they have some familiarity with. Then we can progress to more granular details. In chapter 6 we will explore the world of cellular neuroscience. It is there that issues regarding neurotransmitters, potentials, and similar details will be explored.

The human brain consists of approximately 100 billion neurons and a total of 100 trillion connections,[1] weighing approximately three pounds (1.4 kilograms). In addition to neurons, there are billions more glial cells, in fact about 10 to 50 times more glial cells than neurons.[2] This complexity of the mammalian brain is hard to overstate. The National Institute of Health describes the brain by saying

> The brain is the most complex part of the human body. This three-pound organ is the seat of intelligence, interpreter of the senses, initiator of body movement, and controller of behavior. Lying in its bony shell and washed by protective fluid, the brain is the source of all the qualities that define our humanity. The brain is the crown jewel of the human body.[3]

NEUROANATOMY

Clearly, a single chapter in a single book cannot cover all the aspects of neuroanatomy. Entire books, rather lengthy books, have been written on this topic. The goal of this chapter is to provide the reader a general overview of key elements in neuroanatomy and neurophysiology. For the reader without a background in neuroscience, this should be sufficient to facilitate working with machine learning related to neuroscience. For other readers, this chapter can aid in reviewing key elements of neuroanatomy and neurophysiology.

Before we can begin to explore (or for some readers review) the functionality of the brain, it is necessary to begin with understanding the areas of the brain. Neuroanatomy is a necessary precursor to understanding neurophysiology, cognitive neuroscience, or almost any other area of neuroscience one cares to consider. The basics of neuroanatomy will be covered in this section, from different perspectives.

DOI: 10.1201/9781003230588-7

That means that the same anatomical region will, at least in some instances, be discussed more than once. Viewing the same subject from different perspectives can aid in learning that subject.

NEUROSCIENCE TERMINOLOGY

Before we delve into various views of anatomy and physiology, it will be necessary to become acquainted with certain terminology that is used in neuroscience. Understanding this terminology will aid in reading the rest of this chapter. The following is a basic glossary to aid you in the rest of this chapter.
 Afferent nerve fiber: Originating from the present point.

 Anterior: In front of; toward the face.
 Caudal: Toward the back of the brain or the bottom of the spinal cord.
 Commissure: As bilateral connection of axons connecting the left and right side of the same brain region.
 Decussation: Nerve fibers that cross the sagittal plane from one side of the central nervous system to the other.
 Dorsal: Toward the top of the brain or the back of the spinal cord.
 Efferent nerve fiber: Arriving at the present point.
 Fissure: A deep groove in the brain.
 Fovia: A pit or depression, e.g. the fovia centralis in the center of the macula lutea of the retina.
 Ganglion: A group of neurons usually in the peripheral nervous system.
 Gyrus: An outward folding of the brain.
 Inferior: Below; toward the feet.
 Interstitial: Within spaces.
 Lateral: Toward the edge.
 Medial: Toward the middle.
 Posterior: Behind; toward the back.
 Rostral: Toward the front of the brain or the top of the spinal cord.
 Superior: Above; toward the head.
 Tectum: The dorsal portion of the midbrain.
 Tegmentum: The ventral portion of the midbrain.
 Ventral: Toward the bottom of the brain or the front of the spinal cord.

Other terms will be introduced as needed, in both the current chapter and in subsequent chapters. However, the preceding lists of terms are fundamental to neuroscience, and you should be familiar with them.

DEVELOPMENT VIEW

Neuroanatomy is often considered from a perspective of developmental regions. By developmental perspective, it is meant to examine which regions of the brain evolved first. This begins with the rhombencephalon, or hindbrain. This is the oldest area of the brain and evolved in the last common ancestor of arthropods and chordates approximately 560 million years ago. This region of the brain includes two subregions, the metencephalon and myelencephalon. After a period of development, the rhombencephalon of a fetus will consist of the medulla, pons, and cerebellum. The rhombencephalon is responsible for fundamental physiological functions, such as respiratory rhythm, motor activity, etc.

As was mentioned earlier, the rhombencephalon will develop into the medulla, pons, and cerebellum. Each of these have specific uses. The medulla transmits signals between the spinal cord and the rest of the brain. The medulla also controls autonomic functions such as respiration and heartbeat. The cerebellum is involved in motor control. The cerebellum also plays a role in attention, language, and emotion. Perhaps the most notable feature of the cerebellum, anatomically speaking, is that in the human brain the cerebellum appears much like a structure added on to the brain at the bottom. Figure 5.1 is an image of a human brain oriented left to right. At the bottom of the brain is what appears to be a separate structure that is the cerebellum. While the cerebellum may, at first glance, appear to be a separate structure, it is not.

The pons is an interesting region of the brain. The name pons is Latin for bridge. The pons is sometimes called pons Varolii in honor of the Italian anatomist Costanzo Varolio. The pons is part of the brainstem. The pons is involved in equilibrium, taste, hearing, respiration, and other involuntary actions. Some studies suggest the pons may be related to dreams and even sleep paralysis. The pons will be discussed in more detail in the section of this chapter on *anatomical view*.

The next region of the brain, from an evolutionary perspective, is the mesencephalon or midbrain. This portion of the brain is associated with motor control, arousal, temperature regulation, hearing, and vision. The mesencephalon is difficult to see from the exterior of the brain. It is at the top of the brainstem. Figure 5.2 depicts a cross section of the human brain pointing to the mesencephalon in relationship to the pons and the medulla.

The mesencephalon is itself subdivided into the cerebral aqueduct, cerebellar peduncles, tegmentum, and tectum. As the name suggests, the cerebral aqueduct is a conduit for fluid, in this case cerebrospinal fluid (CSF). The cerebral aqueduct connects the third and fourth ventricles, allowing the cerebrospinal fluid to move between ventricles. The cerebellar peduncles are two stalks, which attach the cerebrum to the brainstem. These are located on either side of the mesencephalon.

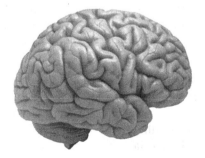

FIGURE 5.1 Cerebellum.

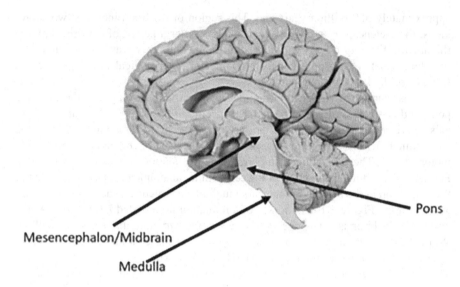

Pons

Mesencephalon/Midbrain

Medulla

FIGURE 5.2   Mesencephalon.

TABLE 5.1
Cranial Nerves

Cranial nerve 1	Olfactory nerve (CN I)
Cranial nerve 2	Optic nerve (CN II)
Cranial nerve 3	Oculomotor nerve (CN III)
Cranial nerve 4	Trochlear nerve (CN IV)
Cranial nerve 5	Trigeminal nerve (CN V)
Cranial nerve 6	Abducens nerve (CN VI)
Cranial nerve 7	Facial nerve (CN VII)
Cranial nerve 8	Vestibulocochlear nerve (CN VIII)
Cranial nerve 9	Glossopharyngeal nerve (CN IX)
Cranial nerve 10	Vagus nerve (CN X)
Cranial nerve 11	(Spinal) Accessory nerve (CN XI)
Cranial nerve 12	Hypoglossal nerve (CN XII)

The tegmentum and tecum are at the ventral and dorsal area of the mesencephalon, respectively. Thus, the tectum forms the top of the mesencephalon while the tegmentum forms the bottom. The tegmentum contains the reticular formation, cranial nerves, the nuclei associated with the cranial nerves, as well as other nuclei. All 12 cranial nerves are summarized in Table 5.1.

The tectum is the dorsal portion of the midbrain and is involved in visual and auditory reflexes. The tectum can be spatially contrasted with the tegmentum, which is the ventral part of the midbrain. The tegmentum is the location of several cranial nerve nuclei.

The term nuclei, in the context of neuroanatomy, may be new to some readers. Throughout the brain and spinal cord, there are groupings of neurons with common functions; these are the nuclei. They are found throughout the spinal cord (ventral and dorsal horn), brain stem (cranial nerve nuclei, reticular formation) and diencephalon (nuclei of thalamus, hypothalamus, subthalamus, and metathalamus), basal ganglia (caudate, putamen, globus pallidus, substantia nigra), and in the cerebral cortex (amygdaloid nuclei).

Another important structure in the tegmentum is the red nucleus. You may also hear the red nucleus referred to as the nucleus ruber. In vertebrates that lack a substantial corticospinal tract, the red nucleus is responsible for controlling gait. However, in other mammals, including primates, the red nucleus is not as important and may be considered partially vestigial.

The prosencephalon, or forebrain, is the third area of the brain. This area is quite developed in certain mammals, including primates. It is responsible for body temperature, reproductive functions, emotions, cognitive functions, and is involved in sleeping and eating as well. This is the area of the brain that contains the cerebral hemispheres. The prosencephalon also is home to the thalamus, hypothalamus, epithalamus, and subthalamus. The thalamus is central for relaying between the medulla oblongata and the cerebrum. The hypothalamus is related to hunger, thirst, body temperature, blood pressure, and even pleasure/pain.

Medulla Oblongata (myelencephalon is also known as the medulla). The medulla lies between the pons rostrally and the spinal cord caudally. It is continuous with the spinal cord just above to foramen magnum and the first spinal nerve. The posterior surface of the medulla forms the caudal half of the fourth ventricle floor, and the cerebellum, its roof. The base of the medulla is formed by the pyramidal descending fibers from the cerebral cortex. The medulla tegmentum contains ascending and descending fibers and nuclei from the 9th (glossopharyngeal), 10th (vagus), 11th (accessory), and the 12th (hypoglossal) nerves. The corticospinal fibers (pyramid) are alongside the anterior median fissure, and decussate (cross the midline) to the contralateral side on their way to the spinal cord. Other prominent structures in the medulla are the inferior olive, and the inferior cerebellar peduncle. The medulla contains nuclei which regulate respiration, swallowing, sweating, gastric secretion, cardiac, and vasomotor activity.

The diencephalon is the most rostral structure of the brain stem; it is embedded in the inferior aspect of the cerebrum. The posterior commissure is the junctional landmark between the diencephalon and the mesencephalon. Caudally, the diencephalon is continuous with the tegmentum of the midbrain. During development, the diencephalon differentiates into four regions: thalamus, hypothalamus, subthalamus, and epithalamus.

ANATOMICAL VIEW

Next, we will review major anatomical regions from a purely anatomical perspective, rather than the evolutionary/developmental perspective. This will necessitate covering the same regions again, but from a different perspective. Approaching the same topic from different perspectives is often an effective way to evaluate new material (assuming basic neuroanatomy is new to you). Figure 5.3 shows a general overview of major regions of the human brain.

In addition to understanding the gross regions of the brain, it will be important to understand directional terms used. Figure 5.4 illustrates many of these terms. Note, some of these terms were introduced in the beginning of this chapter. However, it is often helpful to see these directional terms shown graphically.

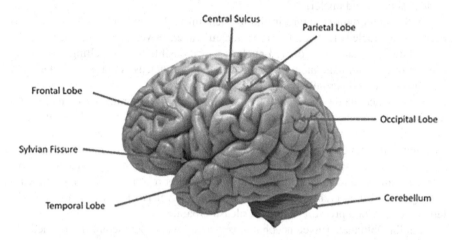

FIGURE 5.3   General anatomical regions.

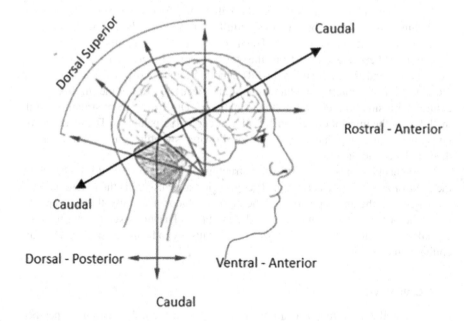

FIGURE 5.4   Anatomical directions.

Generally speaking, dorsal or ventral refers to the upper side. Rostral refers to the front of the body and caudal to the rear. Anterior generally refers to the front, whereas superior to the upper portion. Medial refers to locations to the midline, and distal to those locations farthest from the point of origin. The location of various anatomical regions is critical to understanding neuroanatomy. So being familiar with these terms is essential to the rest of the neuroscience in this book.

Brainstem

We will begin with the brainstem. The brainstem is not of particular interest in higher brain functions such as cognition and memory. However, it is very important in basic biological functions. The brainstem is the most caudal part of the brain and occupies approximately 2.6% of the mammalian brain weight. It consists of the midbrain (mesencephalon), pons, and medulla oblongata. The cerebellum, pons, and medulla oblongata are often grouped together under the name hindbrain (rhombencephalon). The brainstem regions can be seen in Figure 5.5.

Except for cranial nerve nuclei I and II, the brainstem contains all the cranial nerve nuclei. The brainstem also contains the sympathetic and parasympathetic nuclei. This means that the brainstem is critical to bodily functions such as heart rate and breathing.

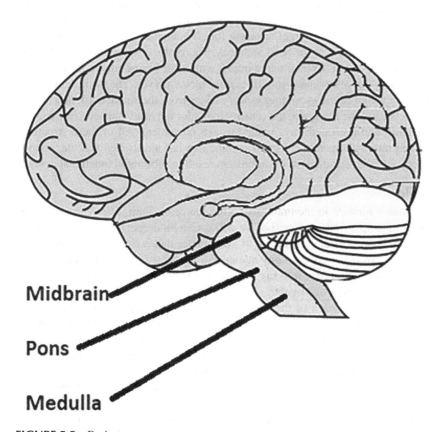

Midbrain

Pons

Medulla

FIGURE 5.5   Brainstem.

The medulla develops from the myelencephalon, and is responsible for a range of autonomic functions. The medulla oblongata was referred to as "the bulb" in the past. This is the origin of terms such as bulbar (i.e., bulbar palsy). The term bulbar, in reference to neuroanatomy, refers to the medulla oblongata. Various regions of the medulla oblongata are responsible for diverse physiological functions. As one example, the ventrolateral medulla plays a role in regulating arterial blood pressure. This is done via sympathetic nerves that target the heart and peripheral blood vessels. Another example is the rostral ventrolateral medulla, which is responsible for sympathetic nervous activity related to cardiac function.

The midbrain is further divided into the tectum, ventral tegmental area, and tegmentum. The tectum is involved in reflex responses, particularly those to visual or auditory stimuli. The ventral tegmental area is a group of neurons on the midline of the dorsal portion of the midbrain. The tegmentum is the ventral region of the midbrain.

The pons is between the medulla oblongata and the midbrain, and contains neurological tracts that carry signals to other brain regions such as the cerebrum and cerebellum. The pons is also involved in coordinating activities of the two cerebral hemispheres. The medulla oblongata is the lower portion of the brainstem and is contiguous with the spinal cord. A number of important involuntary functions include nerve tracts through the medulla, including respiration, heart rate, and blood pressure.

Understanding the brainstem is widely regarded as critical for clinical medicine. The brainstem is either the target or origin for all cranial nerves that deal with sensory and motor function in the head and neck. Therefore, this neurological structure is critical to movement. Cranial nerve nuclei within the brainstem are the targets of cranial sensory nerves or the origin of cranial motor nerves. The brainstem is also the pathway for all of the ascending sensory tracts from the spinal cord, descending motor tracts from the forebrain, the sensory tracts for the head and neck (the trigeminal system), and even pathways related to eye movement. All of these structures are located into a comparatively small volume that has a restricted vascular supply. Therefore, vascular accidents in the brainstem result in substantial functional deficits.

Cerebellum

We next move forward in the brain, exploring the cerebellum. The cerebellum was discussed in the previous section, but we will dive a bit deeper in this section. Most of the cerebellum is actually made up of gray matter that is tightly folded into what is termed the cerebellar cortex. Gray matter will be discussed in more detail in later chapters, but for now suffice it to say that the two major types of matter one finds in the brain are gray and white matter.

To give some indication of the extensive folding in the cerebellar cortex, it has been estimated that if one would completely unfold it and lay it out flat, it would be about one meter long and five centimeters wide. Underneath that layer of gray matter is a mass of myelinated nerve fibers, white matter. Within the cerebellum are four specific deep cerebellar nuclei: dentate, emboliform, fastigii, and globose. A nucleus is simply a cluster of neurons in the central nervous system. As we explore neuroanatomy, we will encounter several important nuclei.

The cerebellum has two hemispheres that are connected via a midline mass called the vermis. The cerebellum has three lobes: the anterior, posterior, and

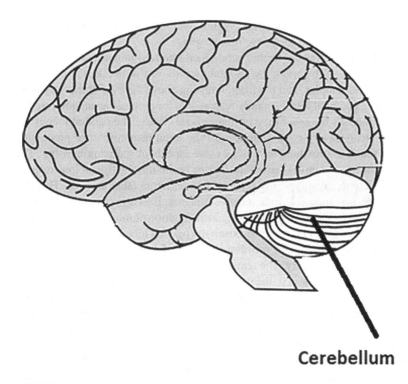

Cerebellum

FIGURE 5.6   Cerebellum.

flocculonodular. The anterior lobe is important in mediating unconscious proprioception. Proprioception is the sense of body position and movement. The posterior lobe is significantly larger than the anterior lobe and is responsible for fine motor movement and coordination. The flocculonodular lobe is involved in regulating eye movement and is thus involved in visual tracking. The cerebellum was shown in earlier figures, but is pointed out here in figure 5.6.

Cerebrum

The cerebrum is perhaps the area of most interest in fields such as cognitive science and psychology. The cerebrum is involved in processing sensory information, language, communication, learning, memory, and a wide range of functions considered "higher" functions.

The cerebral cortex is divided into two hemispheres, just as the previously discussed cerebellum. These two sides of the brain are connected by the corpus callosum, a bridge of wide, flat neural fibers that help relay signals between them. While several popular books suggest these two sides of the brain are specific to particular brain functions, many cognitive tasks are seen across both hemispheres. The separation of brain function by the sides of the brain is called lateralization. However, as was stated, many cognitive functions are not specific to a given hemisphere. One notable exception is language. Two key areas involved in language reside on the left side of the brain. These two areas are Broca's area and Wernicke's area. Broca's area is responsible for language grammar and syntax. Wernicke's area is involved in language content and processing meaning from language.

The cerebral hemispheres are further subdivided into four major lobes: the occipital, towards the back of the brain; the parietal, just above the ear; the temporal, just behind the forehead temples; and the frontal, resting above the eyes at the very front of the cortex. The occipital lobe is mainly responsible for processing and interpreting visual information. It's the seat of the primary visual cortex. The parietal lobe is the home of the somatosensory cortex, the area of the brain responsible for processing sensation and touch information, as well as some aspects of spatial processing. The frontal lobe is the most complex part of the human brain. It is the frontal lobe responsible for those activites that typically separate humans from other animals. This, the largest brain lobe, is responsible for executive function. Executive functions are generally defined as decision-making, planning, and execution of movement. The frontal lobe is the most anterior, and is separated from the parietal lobe by the central sulcus.

The temporal lobe is the major processing center of sound (including language) and also plays a role in memory. The temporal lobe extends almost as far anterior as the frontal lobe but is inferior to it, the two lobes being separated by the lateral fissure (also called the Sylvan fissure). The superior aspect of the temporal lobe contains the cortex, concerned with audition, and inferior portions deal with highly processed visual information. Hidden beneath the frontal and temporal lobes, the insula can be seen only if these two lobes are removed or separated.

The subcortical structures are a group of diverse structures found deep within the brain. They include the diencephalon (thalamus, epithalamus, subthalamus, and hypothalamus), pituitary gland, limbic structures, and the basal ganglia. One important example is the hypothalamus. The hypothalamus and pituitary gland are involved in hormone production and regulation.

The cerebrum is involved in processing sensory information, language, communication, learning, memory, and a wide range of functions considered "higher" functions. The cerebrum lobes are shown in Figure 5.7.

The cerebral cortex is the outer layer of the cerebrum in mammals. Most of the cerebral cortex consists of the neocortex. Approximately 10% of the cerebral cortex is made up of the allocortex. The allocortex consists of three to four layers of neurons. The neocortex consists of six layers labeled from I to VI. It should be noted that while the neocortex in more developed mammals such as dolphins and primates has ridges and grooves (gyri and sulci), it is smooth in smaller mammals.

Limbic System

The limbic system is a very important part of the brain; it is located immediately below the medial temporal lobe of the cerebrum. The limbic system is involved in emotional responses as well as long-term memory. The word limbic originates from the Latin word *limbus*, which means border. As the name *system* suggests, this region of the brain is a system of components. The limbic system was first defined by neuroscientist Paul MacLean. The limbic system can be divided into cortical and subcortical areas. The cortical areas include the limbic lobe, entorhinal cortex, orbitofrontal cortex, piriform cortex, and fornix. The subcortical areas include the hippocampus, amygdala, septal nuclei, and nucleus accumbens. The limbic system is shown in figure 5.8.

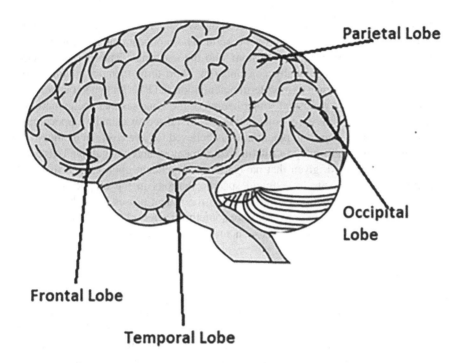

FIGURE 5.7 Cerebrum.

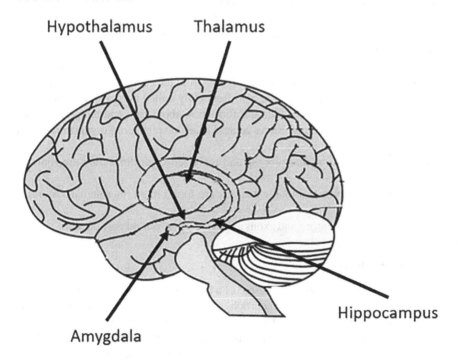

FIGURE 5.8 The limbic system.

The hippocampus is a vital part of the limbic system. There are actually two hippocampi, one in each hemisphere of the brain. The hippocampus is located in the allocortex, but in primates has extensions into the neocortex. The hippocampus is involved in memory, specifically consolidating short-term memory to long-term memory. An activity known as long-term potentiation (LTP) is known to occur in the hippocampus. LTP is a persistent strengthening of the synapses that occurs between two neurons.[4] In fact, long-term potentiation was first discovered in the hippocampus of rabbits in 1966.[5] Long-term potentiation is believed to have a role in memory.

The hippocampus also plays a role in spatial memory and navigation. These two activities are connected, given that navigation is dependent upon special memory. Experiments with rodents have found specific neurons in the hippocampus that respond with bursts of action potentials when the rodent passes through a specific area in its environment. These neurons are sometimes referred to as place cells. The area of the environment that induces such a reaction is referred to as a place field.

SPINAL CORD

The interior of the spinal cord consists of neurons/gray matter while the exterior is white matter. In transverse sections of the spinal cord, the gray matter is usually divided into dorsal (posterior) lateral and ventral (anterior) "horns." The neurons of the dorsal horns receive sensory information that enters the spinal cord via the dorsal roots of the spinal nerves. The ventral horns contain the cell bodies of motor neurons that send axons via the ventral roots of the spinal nerves to striated muscles. The white matter of the spinal cord is subdivided into dorsal (or posterior), lateral, and ventral (or anterior) columns, each of which contains axon tracts related to specific functions.

The dorsal columns bring sensory information from somatic mechanoreceptors. The lateral columns include axons that traverse from the cerebral cortex to the spinal motor neurons. These pathways are also frequently called corticospinal tracts. The ventral columns carry both ascending information about pain and temperature, and descending motor information. Some general rules of spinal cord organization are:

1. Neurons and axons that process and relay sensory information are found dorsally;
2. Preganglionic visceral motor neurons are found in an intermediate/lateral region;
3. Somatic motor neurons and axons are found in the ventral portion of the cord.

NEUROPHYSIOLOGY

Anatomy is concerned with the structure of a biological system, whereas physiology is focused on the function. Thus, neurophysiology is focused on the function of the nervous system. Historically, electrical activity has been a primary way of examining neurophysiology. The history of neurophysiology is extensive. In 177 Claudius Galenus (some sources report his name as Aelius Galenus), often simply called Galen, posited that the brain was the center of thought. This was contrary to the belief that the heart was the seat of thoughts and emotions. No less a personage than Aristotle had championed the heart as the center of human intellect.

Study into neurophysiology largely languished for many centuries. In the mid 1500s, Italian physician Nicolo Massa described the effects of diseases on the nervous systems. He also noted the presence of cerebrospinal fluid in ventricles. By the 1700s there began to be a substantial increase in studies of the brain. During this time, Luigi Galvani described electrical activity in the nerves of dissected frogs. Also, during the 1700s, Jean-Baptiste Le Roy began using electroconvulsive therapy on mentally ill patients. By the 1800s, more modern understandings began to emerge. During this century, Theodor Schwann discovered the myelin sheath.

It was also during the 1800s that perhaps the most famous clinical case in neuroscience occurred. In 1848, Phineas Gage was in an accident that led to an iron rod piercing his brain. That accident destroyed a large part of the left side of his frontal lobe. This caused substantial effects on his personality and behavior. While Mr. Gage lived for 12 years after the accident, many of his friends and associates stated that he was "no longer Gage." His case strongly suggests the correlation between neurophysiology and personality.

Neurophysiology has traditionally been dominated by electrophysiology, using EEG to study neurofunction. In recent decades, this has expanded to additional imaging technologies. New technologies have provided increased insight into how the brain functions. New data has also shown the role that various neurotransmitters play.

NEUROTRANSMITTERS

Neurotransmitters are chemicals that cross the synaptic cleft between the transmitting neuron and the receiving neuron. This transmission is caused by the electrical impulse in the sending neuron's axon. Neurons behave according to Dale's law (sometimes referred to as Dale's principle) which states that a neuron will perform the same chemical activity at all of its synaptic connection to other cells. This activity is independent of the identity and nature of the target cells. This principle is attributed to Sir Henry Hallett Dale, a physiology researcher of the early 20th century. He received the 1936 Nobel Prize in Physiology or Medicine for his work. The principle basically means that when a neuron is excited, all of its axonal branches will secrete the same neurotransmitter(s).

METABOLISM

Metabolism is the method of processing chemicals by the body. Vertebrates have a blood-brain barrier that blocks some chemicals from crossing and being metabolized by neurons in the brain. The blood-brain barrier is made up of endothelial cells and is a semipermeable border. The blood-brain barrier restricts the passage of large or hydrophilic molecules while permitting the diffusion of hydrophobic molecules.

Metabolism is an important aspect of neurophysiology. The brain consumes a disproportionate amount of energy. A substantial part of the brain's energy consumption is devoted to maintaining the membrane potential of neurons. Vertebrates in general use 2% to 8% of their basal metabolism for the brain, while primates have a higher percentage. In humans, the brain can use 20% to 25% of the basal metabolism. Most energy for the brain comes from the metabolism of glucose, though there can be contributions from fatty acids, lactate, and other chemicals.

NEUROIMAGING

One of the primary ways to study the brain, whether the goal is anatomical or physiological, is via neuroimaging. Electroencephalography has been a primary means of studying brain activity. This is due to two factors, the first being the fact that EEG is noninvasive, thus easier on test subjects. The second factor is that EEG is less expensive than many other imaging techniques, such as functional magnetic resonance imaging (fMRI), functional near infrared spectroscopy (fNIR), and other methods.

Electroencephalography began with the research of Richard Caton, a physician who published findings on electrical phenomena of exposed cerebral hemispheres in lab animals.[6] This work was expanded upon by others. As one example, in 1912 the first results of animal EEG were published by Ukrainian physiologist Vladimir Pravdich-Neminsky.[7] By 1947, the American EEG Society was founded. Work with EEG results has been a major part of research in neurology and psychology since that time. More recently, technologies such as machine learning have been used to analyze EEG results.

Neuroimaging technology now encompasses a number of approaches other than EEG. These technologies include direct measures that detect electrical (e.g., electroencephalography, EEG) or magnetic activity (e.g., magnetoencephalography, MEG) of the brain. Other technologies that depend on indirect measures of brain function reflecting brain metabolism or hemodynamics of the brain are also used. This includes techniques such as functional magnetic resonance imaging (fMRI), functional near-infrared spectroscopy (fNIRS), and positron emission tomography (PET). In addition, direct brain-monitoring approaches typically provide detailed information with high temporal resolution, but they normally lack spatial coverage; indirect measures show higher spatial resolution than direction measures. However, EEG still remains widely used in research due to its low cost and ease of implementation.

NEUROFUNCTION

As was stated earlier in this chapter, neurophysiology is about the function of the nervous system, whereas anatomy is about the structure. Put in more simple terms, neurophysiology is concerned with what the brain does. In this section, we will briefly address the major functions of the nervous system. As with the rest of this book, our primary focus is on human neuroscience.

Motor Control

An important aspect of the nervous system is motor control. There are multiple areas of the brain that connect to the spinal cord and aid in controlling muscles throughout the body. Table 5.2 summarizes the major neurological areas related to motor control.

The cerebellum was discussed earlier in this chapter in some detail. However, the other regions in the preceding table will be briefly described.

The ventral horn is comprised of three columns of gray matter in the spinal cord. The ventral horn contains neurons that affect the skeletal muscle. In contrast, the

TABLE 5.2
Neurological Areas Related to Motor Control

Area	Location	Function
Ventral horn (also called the anterior horn or motor horn)	Spinal cord	Includes motor neurons that activate muscles.
Oculomotor nuclei	Midbrain	Includes motor neurons that activate the ocular muscles.
Cerebellum	Hindbrain	Calibrates precision and timing of movements.
Basal ganglia	Forebrain	Action selection on the basis of motivation.
Motor cortex	Frontal lobe	Direct cortical activation of spinal motor circuits.
Premotor cortex	Frontal lobe	Groups rudimentary movements into organized movement.
Supplementary motor area	Frontal lobe	Sequences movements into temporal patterns.

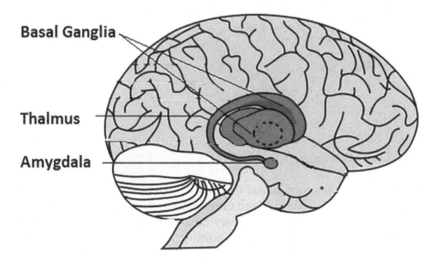

FIGURE 5.9 Basal ganglia.

posterior horn (also called the posterior grey column) contains neurons that receive sensory information. Alpha motor neurons are located in the ventral horn.

The oculomotor nucleus' purpose is given in the name. It controls the ocular muscles. This nucleus is located in the mesencephalon and can be broken onto many smaller segments. A more exact location of the oculomotor nucleus is that it is located in the central midbrain at the level of the superior colliculus ventral and medial periaqueductal gray matter.

The basal ganglia, also called the basal nuclei, are a group of related nuclei that exist in the brains of all vertebrate animals. The basal nuclei are important in control of voluntary motor function. There are other neurological functions that the basal nuclei are involved in; however, voluntary motor function is our focus in this section. Figure 5.9 should aid you in locating the basal ganglia.

Premotor cortex   Primary Motor Cortex

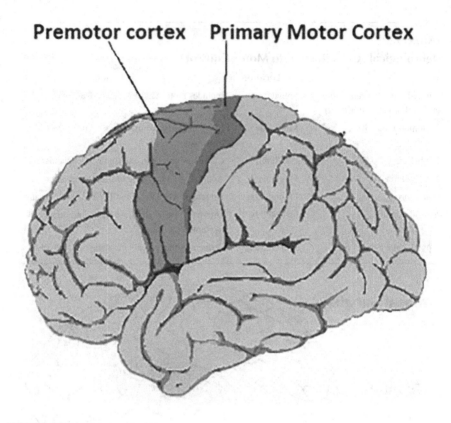

FIGURE 5.10 Motor cortex.

The motor cortex is a subregion of the cerebral cortex. The motor cortex is important for the executive functions related to voluntary motor control. This means planning, control, and execution of voluntary motor activity. The premotor cortex is within the frontal lobe and anterior to the primary motor cortex. The premotor cortex and the primary motor cortex are shown in Figure 5.10.

The supplementary motor area is a region of the motor cortex that is also involved in control of movement. Neurons in the supplementary motor area (SMA) project into the spinal cord and are thus believed to have a direct role in controlling movement.

Perception

Another vital function of the nervous system is that of perception. This means perceiving visual, auditory, and proprioceptor stimuli. The process of perception begins with receptor cells receiving stimulus, and then passing that via nerves to particular areas of the brain. The process of taking input from some portion of the body and processing it along the nervous system, to a particular part of the brain, is a convoluted procedure. It is beyond the scope of this chapter to completely cover perception; however, a few notable examples should aid you in understanding processing sensory input.

The first example is sight. When light reaches photoreceptor cells on the retina, there is a reaction. Mammals have three types of photoreceptors; rods, cones, and

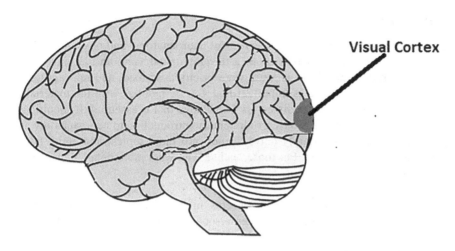

FIGURE 5.11   Visual cortex.

photosensitive retinal ganglion cells. Rods are concentrated on the outer edge of the retina and are important for peripheral vision. Rods are more sensitive to light and thus important for low-light vision, however they have very little involvement in color vision. Cones respond to light of different wavelengths, thus creating color vision. Photosensitive retinal ganglion cells (pRGC) are a type of neuron found in the retina of mammals. These neurons have a light-sensitive protein named melanopsin, thus making them sensitive to light. Photosensitive retinal ganglion cells are involved in the pupillary light reflex, and some studies suggest a role in conscious sight.

Signals from the three photoreceptive types of cells are transmitted by the optic nerve to the central ganglia in the brain. Signals can also travel to the superior colliculus. Ultimately, vision is processed in the visual cortex of the brain. This area of the brain is part of the cerebral cortex and located in the occipital lobe. Both hemispheres of the brain include a visual cortex, which processes visual signals from the opposite side of the body. The visual cortex is highlighted in Figure 5.11.

For those readers who wish to delve deeper into the visual cortex, there are resources available on the internet.[8],[9]

The next example of perception is hearing. Sound waves interact with the ear, particularly the tympanic membrane. The vibrations of the tympanic membrane are passed on to the ossicles (malleus, incus, and stapes), the three smallest bones in the body. Eventually, sound waves pass to the organ of Corti, a portion of the cochlea, and vibrates according to frequency of sound. Sound information is passed along the auditory nerve, to the cochlear nucleus in the brainstem, and then on to the inferior colliculus in the midbrain tectum. The inferior colliculus is responsible for integrating auditory input with other input from other parts of the brain, and is involved in a range of subconscious reflexes. The inferior colliculus also projects the signal to the medial geniculate nucleus located in the thalamus. Information regarding auditory stimuli is then related to the primary auditory cortex, which is found in the temporal lobe.

SUMMARY

This chapter provided a general overview of neuroanatomy and physiology. Neuroanatomy was presented from both a developmental view and an anatomical view. Important regions of the brain were briefly described, and relevant terminology was introduced. It is important that you be familiar with the material in this chapter as it provides rudimentary knowledge that is essential. It should also be noted that this chapter is only a rudimentary introduction to neuroanatomy and physiology. One can certainly delve deeper into these topics, and entire books have been written on them. For those readers who want more detail, consider the following resources:

Presti, David E. Foundational Concepts in Neuroscience: A Brain-Mind Odyssey (Norton Series on Interpersonal Neurobiology) (p. 82). W. W. Norton & Company.
Carpenter, Roger; Reddi, Benjamin. Neurophysiology: A Conceptual Approach, Fifth Edition. CRC Press. Kindle Edition.

TEST YOUR KNOWLEDGE

1. The _____ calibrates precision and timing of movements.

    a. Supplementary motor area

    b. Primary motor cortex

    c. Premotor cortex

    d. Cerebellum

2. The _____ sequences movements into temporal patterns.

    a. Supplementary motor area

    b. Primary motor cortex

    c. Premotor cortex

    d. Cerebellum

3. A nerve that originates from the current point is said to be what?

    a. Anterior

    b. Afferent

    c. Caudal

    d. Efferant

4. What is the term for the dorsal portion of the midbrain?

   a. Tegmentum

   b. Pons

   c. Medulla

   d. Tectum

5. A cluster of neurons in the central nervous system is referred to as a _____.

   a. Gyrus

   b. Nuclei

   c. Ganglion

   d. Lob

6. Which of the following brain regions is most closely associated with processing meaning from language?

   a. Broca's area

   b. Temporal lobe

   c. Occipital Lobe

   d. Wernicke's area

7. Which of the following brain regions is most closely associated with processing visual information?

   a. Parietal Lobe

   b. Occipital Lobe

   c. Temporal Lobe

   d. Broca's area

8. The blood-brain barrier is easily passed by:

   a. Hydrophobic molecules

   b. Hydrophilic molecules

   c. Neurotransmitters

   d. Nothing

9. Explain the Hallet principle.

The answer should be something like "When a neuron is excited, all of its axonal branches will secrete the same neurotransmitter(s)."

NOTES

1. Scientific American. 2011, January. www.scientificamerican.com/article/100-trillion-connections/
2. Herculano-Houzel, S. 2012. The remarkable, yet not extraordinary, human brain as a scaled-up primate brain and its associated cost. *Proceedings of the National Academy of Sciences*, 109(Supplement 1), 10661–10668. www.pnas.org/content/109/Supplement_1/10661
3. www.ninds.nih.gov/health-information/public-education/brain-basics/brain-basics-know-your-brain
4. www.sciencedirect.com/science/article/pii/S0896627316309576
5. https://physoc.onlinelibrary.wiley.com/doi/abs/10.1113/jphysiol.1973.sp010273
6. İnce, R., Adanır, S.S. and Sevmez, F. 2021. The inventor of electroencephalography (EEG): Hans Berger (1873–1941). *Child's Nervous System*, 37(9), 2723–2724.
7. Nedvědová, M. and Marek, J. 2018. Comparing EEG signals and emotions provoked by images with different aesthetic variables using emotive insight and neurosky mindwave. In *17th Conference on Applied Mathematics APLIMAT 2018: Proceedings*. Slovenská Technická Univezita v Bratislave.
8. www.ncbi.nlm.nih.gov/books/NBK482504/
9. www.sciencedirect.com/topics/neuroscience/visual-cortex

6 Cellular Neuroscience

INTRODUCTION

In chapter 5, general neurological anatomy and physiology were covered. In this chapter, the topic of cellular neuroscience will be explored. While physiology provides an understanding of the functionality of the brain, many issues in neuroscience require one to take a cellular or even molecular view. All neurological activity begins with chemical movement across the neuron's membrane to begin a signal process that will propagate to one or more neighboring neurons.

BASIC NEURO CELLULAR STRUCTURE

Neurons are specialized cells. As with other cells, they have a nucleus, ribosomes, endoplasmic reticulum, and other common cellular organelles. The focus in this chapter will be on those aspects of neurons that are not common to the other cells in a mammalian body. Figure 6.1 shows the general structure of a neuron.

Mammalian cells, including neurons, can be divided into three primary regions. There is the cell membrane, the cytoplasm, and the nucleus. The nucleus of a neuron

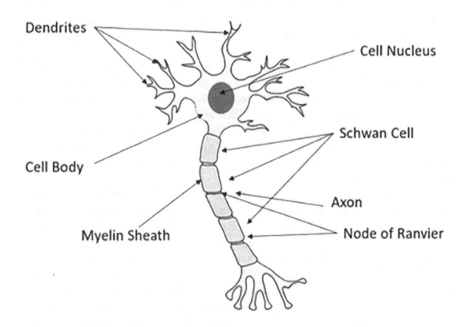

FIGURE 6.1 Neuron structure.

DOI: 10.1201/9781003230588-8

is essentially the same as most mammalian cells. This is where the genetic material for the cell is stored. For the purposes of this book, detailed analysis of the cell's DNA won't be a focus.

In addition to organelles common to all mammalian cells, neurons have numerous Nissl bodies. Nissl bodies (sometimes called Nissl granules or Nissl substance) are made of rough endoplasmic reticulum and are named after neuroscientist Franz Nissl. Endoplasmic reticulum is found in all eukaryotic cells. The rough endoplasmic reticulum is studded with ribosomes for protein manufacture. These Nissl structures are related to the abundant production of proteins in the neuron.

Neurons have a mesh of neurofilaments that support the cell body structure.

The cell membrane is where molecules are brought into the cell or expelled from the cell. As with all mammalian cells, mechanisms to facilitate transport of specific molecules across the cell membrane are critical. However, with neurons this process takes on even more importance. When certain ions cross the membrane, electrical current is continued, eventually leading to a signal down the neuron's axons, leading to passing neurochemicals across the synapse to the next neuron's dendrites. This is the foundation of neurological signaling, and in turn the basis for all brain activity. This will be explored in greater detail later in this chapter.

The primary chemical building blocks of cell membranes are phospholipids. This is true for neurons as it is for all cells. Phospholipids, as with all lipids, contain long nonpolar chains of carbon atoms bonded to hydrogen atoms. In addition to the long chain of carbon atoms, the phospholipid has a polar phosphate group. A phosphate group is a phosphorus atom bonded to three oxygen atoms. This phosphate group is attached to one end of the molecule in a phospholipid.

The cytoplasm or cell body is much like the soma or body of any mammalian cells. It contains typical organelles such as rough endoplasmic reticulum, Golgi apparatus, ribosomes, etc. However, it also has additional appendages that other cells don't have. These are the axon and the dendrites. Dendrites receive signals from other neurons whereas the axon transmits signals from the current neuron to the next neuron. To facilitate the transmission of electrical current down the axon, the axon is encased in a myelin sheath.

The myelin sheath is made up of special cells called Schwan cells. These cells derive their name from the physiologist Theodor Schwann. Schwann cells are a type of glial cell. More details on glial cells a bit later in this section. For now, the focus is on the Schwann cells' function in creating a myelin sheath around the axon. The myelin sheath insulates that axon and thus increases the efficiency of electrical signals passed along the axon.

Along the axon one will note spots that do not have sheathing. These are called nodes of Ranvier. These are quite small, usually about one-micron regions. Schwann cells do have additional functionality, but for now our focus is on the forming of the myelin sheath on the axon. To illustrate the importance of these Schwann cells, consider neuropathology such as multiple sclerosis, which is a disease that degrades the myelin sheath of the nervous system. The effects of multiple sclerosis can be quite severe. Neurological disorders will be examined in more detail in chapter 7.

At the end of the axon is the axon terminal. This contains synapses. Specifically, synaptic boutons are specialized structures that release neurotransmitters into the synaptic cleft so they can reach the next neuron's dendrites. In addition to the synaptic bouton and the axon terminal, one may also find boutons along the length of the axon. These boutons are referred to as en passant boutons.

Dendrites are extensions of the neuron that can receive neurotransmitters that are released from neighboring neuron axons. Dendrites tend to be shorter than axons but have an enlarged surface area to allow them to receive neurotransmitters from neighboring axons. There are also dendrites that receive signals from other dendrites. This signaling process is referred to as dendrodendritic. Neurons can and do grow additional dendrites, creating new connections. This process is called dendritic branching, sometimes also referred to as dendritic arborization.

Water is the primary ingredient of the fluid found inside the neuron. The fluid itself is often referred to as intracellular fluid or cytosol. This is also true of the external environment of the neuron. Ions are dissolved in this water, and they are responsible for the resting and action potentials.

Extending from the body of the neuron are axons and dendrites. Dendrites are fibrous cellular branches that extend out from the cell body. The dendrites receive and process signals from the axons of other neurons. More specifically, dendrites will receive neurotransmitters from the axons of other cells. Neurons will often have multiple sets of dendrites. These sets of dendrites are called dendritic trees. Axons are long extensions that proceed from the cell body at a specialized junction called the axon hillock. Many axons are insulated with a fatty substance called myelin. The myelin insulation improves conductivity of electrical impulses down the axon.

Neurons are of three primary types. There are sensory neurons that respond to stimuli including visual, auditory, and touch stimuli. Next there are motor neurons that receive signals and initiate muscle contractions. Third there are interneurons that connect the sensory neurons to the motor neurons.

TYPES OF NEURONS

There are a number of different types of neurons. Multipolar neurons are one type of neuron. These types of neurons have a single axon and symmetrical dendrites that extend from it. Multipolar neurons are the most common type of neuron found in the central nervous system.

Pyramidal neurons are the largest neuron cells and are primarily found in the cerebral cortex, hippocampus, and amygdala. These neurons have one axon, but several dendrites, to form a pyramid-type shape.

Purkinje neurons are inhibitory neurons. These neurons release neurotransmitters that can inhibit other neurons from firing. Purkinje neurons have multiple dendrites that spread out from the cell body. These neurons are primarily found in the cerebellum and are named after their discoverer Jan Evangelista Purkyně, who first described Purkinje neurons in 1839.[1]

Bipolar neurons are another type of neuron. These neurons have two extensions extending from the cell body. The axon is on one side and the dendrites on the other, thus the term bipolar. Bipolar neurons are mostly found in the retina of the eye and won't be a focus of this book. Unipolar neurons are only found in invertebrate species and have a single axon. We won't be discussing these in this book, but they are listed here just for completeness.

SYNAPSE

One focus of neuroscience is the synapse. This is the small gap between the axon of one neuron and the dendrite of the next. There are actually two types of synapse: chemical and electrical. For mammals, including humans, almost all synapses are chemical. This means that a chemical moves from one neuron to the next across the synapse. In response to the electrical signal propagated down the axon of the sending neuron, neurotransmitters are released into the synapse and taken up by the receiving neuron dendrite. A schematic of the synapse is shown in Figure 6.2.

The synapse, often called the synaptic cleft, is where neurotransmitters move from one neuron to another. We will be examining neurotransmitters in detail later in this chapter. In a chemical synapse, the terminal of the axon (i.e., the presynaptic fiber) provides a chemical connection to the dendrite (i.e., the postsynaptic fiber). This synaptic space, or synaptic cleft, is microscopic, usually one to three microns wide, but can be smaller. The arrival of a nerve impulse at the presynaptic terminals (the axon) causes the synaptic vesicles to move toward the presynaptic. These vesicles will fuse with the membrane and release a neurotransmitter into the synaptic cleft. These vesicles differ in size and shape and can be agranular, spherical, or flattened.

Synapses can also be classified by what is being connected. While an axon-to-dendrite connection is the most common (also called axodendritic synapse), there can be synapses between dendrites of two neurons. A dendrite-to-dendrite synapse is referred to as a dendrodendritic synapse. Dendrodendritic synapses were first discovered by Wilfrid Rall, G.M., Shepherd, T.S. Reese, and M.W. Brightman in 1966.[2] There are also axo-axonic synapses, which consist of an axon terminating on another axon or axon terminal.[3] In addition, there are axodendrosomatic synapses, which occur between the axon of one neuron and the dendrites and cell body of another. There are also axosomatic synapses, which are between the axon of one neuron and the body of another. It should be remembered that the axodendritic synapse is, by far, the most common. However, other synapses do exist and you will encounter them when reading neuroscience literature.

Synapse

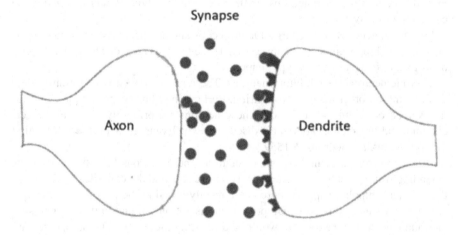

FIGURE 6.2 Synapse.

Electrical Synapses

Chemical synapses are the majority of synapses in mammals. However, there are electrical synapses found in all nervous systems, including the human brain. In a chemical synapse, a neurotransmitter is the communication vehicle between the two neurons. In an electrical synapse, the membranes of the two neurons come particularly close and are linked together by an intercellular specialization called a gap junction. Gap junctions consist of precisely aligned, paired channels in the membrane of the pre- and postsynaptic neurons. Each channel pair forms a pore, and as a result a variety of substances can diffuse between the cytoplasm of the pre- and post-synaptic neurons. This can include not only ions, but molecules as large as several hundred daltons. For those readers unfamiliar with this unit of measurement, a dalton is defined as one-twelfth of the mass of an unbound neutral carbon 12 atom. This allows complex molecules such as ATP and other important intracellular metabolites to be transmitted from one neuron to another.

ION CHANNELS

As was discussed earlier in this chapter, the membrane of a neuron is integral to the process of signaling. Specifically, ion channels are the starting point for neurological signals. These channels are selectively permeable to specific ions such as calcium, sodium, or potassium. Ion transporters and ion channels are responsible for ionic movements across the membranes of neurons. Transporters create ion concentration differences by actively transporting ions against their chemical gradients. Ion channels then take advantage of these concentrations.

Electrical potentials are generated across the membranes of neurons because there are variations in the concentrations of specific ions across nerve cell membranes, and the membranes are selectively permeable to some of these ions. These two facts are in turn due to two different kinds of proteins in the cell membrane. The ion concentration gradients are established by proteins known as active transporters. As the name suggests, active transporters actively move ions into or out of cells against their concentration gradients. The selective permeability of membranes is primarily due to proteins that allow only certain kinds of ions to cross the membrane in the direction of their concentration gradients. These proteins are called ion channels.

To better understand the function of ion gradients and selective permeability in generating a membrane potential, consider a simple example in which a membrane divides two compartments, each containing solutions of ions. In this hypothetical system, it is quite simple to determine the composition of the two solutions and then to control the ion gradients across the membrane. As one example, consider a membrane that is only permeable to potassium ions (K+). If the concentration of K+ on both sides of this membrane is equal, then no electrical potential will be measured across that membrane. However, if the concentration of K+ is not the same on the two sides, then an electrical potential will be generated. For example, if the concentration of K+ on one side of the membrane is four times higher than the K+ concentration on the other side, this will cause the electrical potential of the first side to be negative relative to the other side. This difference in electrical potential is generated because the potassium ions flow down their concentration gradient and carry their electrical charge. A constant resting outflow of K+ is therefore responsible for the resting membrane potential.

Na⁺ is another critical ion in neuroscience. Na⁺ is important for the action potential in neurons. Action potentials are instigated as the extracellular concentration of Na⁺ is altered. As the concentration of sodium in the extracellular solution is reduced, the action potentials decrease.

Once an action potential is initiated, the next issue to consider is the propagation of that signal. There is substantial variation in the velocity of the propagation of action potentials. The propagation velocity of the action potentials in nerves can vary from as little as one-tenth of a meter per second up to 100 meters per second. Because electrical signals are the basis of information transfer in the nervous system, it is important to understand these signals in some detail. One issue for neurons is that the axons, which can be rather long, are not good electrical conductors. To offset for this deficiency, neurons have evolved a system that facilitates the conductance of electrical signals over substantial distances in spite of the neurons' intrinsically poor electrical characteristics. The electrical signals produced by this booster system are sometimes referred to as "spikes" or "impulses."

NEUROTRANSMITTERS

Ultimately, the voltage carried down an axon will cause the release of neurotransmitters across the synaptic cleft, reaching the dendrites on the other side. There are a number of different neurotransmitters; each are related to different neurological functions. The most widely found neurotransmitters in all vertebrates, including humans, are glutamate and gamma-aminobutyric acid (GABA). Glutamate is an excitatory neurotransmitter while GABA is predominantly inhibitory.

There are three criteria for identifying a substance as a neurotransmitter (note that some sources split item two into two separate criteria, thus arriving at four criteria):

1. The chemical must be present in or synthesized in the neuron.
2. When the neuron is active, the chemical must be released and produce some response in the target. That response should be the same if the chemical is placed on the target via experiment.
3. There must be a process for removing the chemical from its site of activation after it has completed its action.

This section will explore specific neurotransmitters, both individual neurotransmitters and classes of neurotransmitters. However, before we get to that, there are general classes of neurotransmitters:

Excitatory. Excitatory neurotransmitters induce the neuron to fire. There are a number of excitatory neurotransmitters, including glutamate, epinephrine, and norepinephrine.

Inhibitory. Inhibitory neurotransmitters block or prevent the chemical message propagated further. These are the counters to excitatory neurotransmitters. Examples of inhibitory neurotransmitters include gamma-aminobutyric acid (GABA), glycine, and serotonin.

Modulatory. Modulatory neurotransmitters moderate the effects of other chemical messengers.

These various types of neurotransmitters work together, such as modulatory neurotransmitters affecting excitatory or inhibitory neurotransmitters. The specific neurotransmitters are important to all neurological activities. These neurotransmitters are integral to physiological functions, emotions, memory, and even mental illness.

ACETYLCHOLINE

Acetylcholine, the first neurotransmitter discovered, was originally described as "vagus stuff" by Otto Loewi because of its ability to mimic the electrical stimulation of the vagus nerve. It is now known to be a neurotransmitter at all autonomic ganglia, at many autonomically innervated organs, at the neuromuscular junction, and at many synapses in the central nervous system. In the autonomic nervous system, acetylcholine (ACh) is the neurotransmitter in the preganglionic sympathetic and parasympathetic neurons. Acetylcholine is also the neurotransmitter at the adrenal medulla and serves as the neurotransmitter at all the parasympathetic innervated organs.

Acetylcholine is the neurotransmitter at neuromuscular junctions, at synapses in the ganglia of the visceral motor system, and at a variety of sites within the central nervous system. Whereas a great deal is known about the function of cholinergic transmission at the neuromuscular junction and at ganglionic synapses, the actions of acetylcholine in the central nervous system are not as well understood.

Acetylcholine is synthesized in nerve terminals from the precursors acetyl coenzyme A and choline. This process is catalyzed by choline acetyltransferase (CAT); therefore, the presence of CAT in a neuron is a strong indication that ACh is used by that neuron. Choline is present in plasma at a concentration of about 10 mM, and is taken up into cholinergic neurons by a high-affinity Na+/choline transporter. Approximately 10,000 molecules of acetylcholine are packaged into each vesicle by a vesicular ACh transporter. The chemical structure of acetylcholine is shown in Figure 6.3:

The following quote[4] regarding acetylcholine may help elucidate this neurotransmitter's function for you:

Acetylcholine (Ach) is a neurotransmitter that functions in both the central and peripheral nervous systems. It is a non-monoamine subtype, meaning that it does not contain an amino group connected to an aromatic ring by a

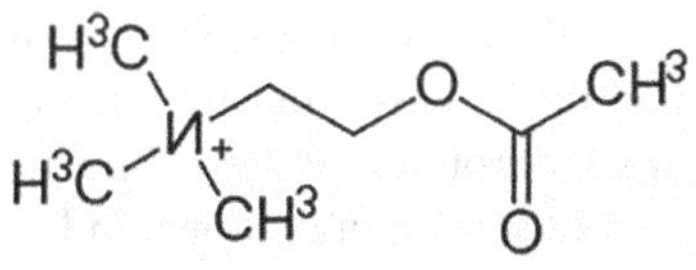

FIGURE 6.3 Acetylcholine structure.

carbon chain (which is common to the neurotransmitters of the noradrenergic, serotonergic, and dopaminergic systems). Instead, it is made up of two chemical groups: choline and acetyl coenzyme A (AcCoA).

Choline is an essential nutrient present in soy, egg yolks, and meat, and is classified within the B-complex group of vitamins. The precursor to choline, AcCoA, is derived from glucose. The synthesis of this neurotransmitter, then, is dependent on adequate consumption of choline; insufficiency can be combatted by taking a choline supplement, usually in the form of lecithin.

CATECHOLAMINES

Dopamine (DA), norepinephrine (also called noradrenaline), and epinephrine (also called adrenaline) are a class of neurotransmitters named on the basis of their chemical structure. More specifically, all of these neurotransmitters have a hydroxylated phenol ring termed a catechol nucleus. These are referred to as catecholamines.

Dopamine (DA) has been called "pleasure chemical" because it is released when mammals receive a reward in response to their behavior. However, the pleasure or reward role of the neurotransmitter has been modified by neuropsychologists. Rather than simply referring to rewards or pleasure, the process is more complex and is referred to as motivational salience. That is the cognitive process of forming attention that motivates the individual either towards or away from a given outcome. In other words, avoidance is as much a part of motivational salience as is seeking pleasure.

Epinephrine (also called adrenaline) and norepinephrine are responsible for the fight-or-flight response stress. Both of these neurotransmitters stimulate your body's response by increasing your heart rate, breathing, blood pressure, blood sugar, and blood flow to your muscles. These neurotransmitters also work to improve attention and focus, thus facilitating quicker reaction time to stressors.

SEROTONIN

Serotonin (5-hydoxytryptamine 5HT), sometimes called the "calming chemical," is best known for its mood-modulating effects. A lack of serotonin has been linked to depression and related neuropsychiatric disorders. But serotonin is farther-reaching, and has also been implicated in helping to manage appetite, sleep, memory, and, most recently, decision-making behaviors.

In the human brain, the primary source of serotonin release is in the raphe nuclei. Raphe nuclei are a cluster of nuclei in the brain stem. There are nine raphe nuclei, labeled B1 to B9.

An interesting fact about serotonin is that, despite its critical role in neuroscience, it is also found in other parts of the human body. In fact, almost 90% of the serotonin in the human body is located in the gastrointestinal tract, more specifically enterochromaffin cells in the GI tract. Serotonin, in the GI tract, regulates intestinal movements. The chemical structure of serotonin is shown in Figure 6.4.

An important aspect of serotonin is an issue called serotonin syndrome. That is described in the following quote from the Mayo clinic:[5]

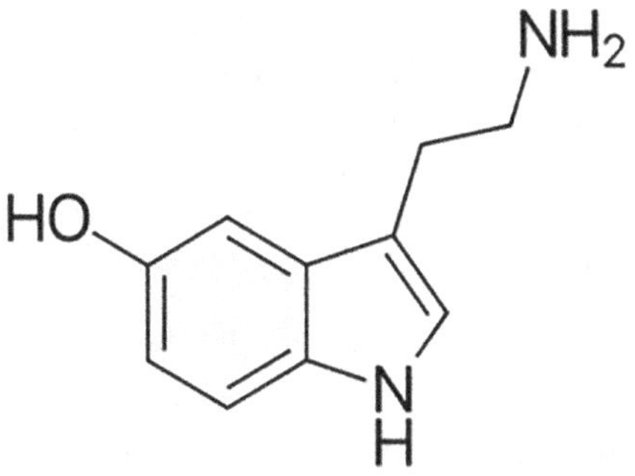

FIGURE 6.4 Serotonin structure.

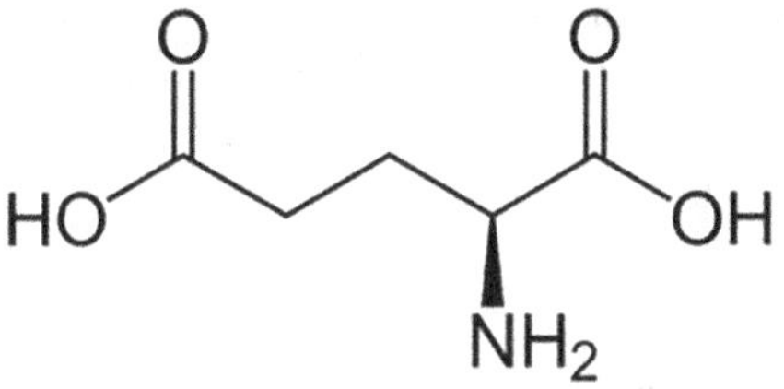

FIGURE 6.5 Glutamate structure.

Serotonin is a chemical that the body produces naturally. It's needed for the nerve cells and brain to function. But too much serotonin causes signs and symptoms that can range from mild (shivering and diarrhea) to severe (muscle rigidity, fever and seizures). Severe serotonin syndrome can cause death if not treated.

Serotonin syndrome can occur when you increase the dose of certain medications or start taking a new drug. It's most often caused by combining medications that contain serotonin, such as a migraine medication and an antidepressant. Some illicit drugs and dietary supplements are associated with serotonin syndrome.

Serotonin syndrome illustrates the important role neurotransmitters play. It also demonstrates that a proper balance of neurotransmitters, and frankly all physiological chemistry, is vital to normal function.

GLUTAMATE

Glutamate (GLU) is the most excitatory neurotransmitter in the cortex. Too much glutamate results in excitotoxicity, or the death of neurons due to stroke, traumatic brain injury, or amyotrophic lateral sclerosis, the debilitating neurodegenerative disorder better known as Lou Gehrig's disease. Yet, it's not all bad news. The excitement GLU brings is important to learning and memory: long-term potentiation (LTP), the molecular process believed to help form memories, occurs in glutamatergic neurons in the hippocampus and cortex. The chemical structure of glutamate is shown in Figure 6.5.

Glutamate is generally acknowledged to be the most important transmitter for normal brain function. Nearly all excitatory neurons in the central nervous system are glutamatergic, and it is estimated that over half of all brain synapses release this agent. Glutamate plays an especially important role in clinical neurology because elevated concentrations of extracellular glutamate, released as a result of neural injury, are toxic to neurons.

INTOLAIMINES

The intolaimines are important in a number of neuropsychological functions. Serotonin (5-hydroxytryptamine; 5-HT) is the principal member of this group of compounds. The name serotonin is derived from the fact that this substance was first isolated from the serum based on its ability to cause an increase in blood pressure. Melatonin, a second indolamine, is restricted to the pineal and is released into the blood stream in a manner that is regulated by the diurnal cycle.

The histamines were not originally recognized as neurotransmitters. Readers are likely familiar with histamines in the context of allergies, and anti-allergy medications (called antihistamines).

GAMMA-AMINOBUTYRIC ACID (GABA)

Most inhibitory neurons in the brain and spinal cord utilize either γ-aminobutyric acid (GABA) or glycine as a neurotransmitter. GABA was identified in brain tissue during the 1950s. David Curtis and Jeffrey Watkins were the first to demonstrate that GABA inhibits the ability of mammalian neurons to fire action potentials. It has since been determined that as many as one-third of the synapses in the brain use GABA as their neurotransmitter. Unlike glutamate, GABA is not an essential metabolite, nor is it incorporated into protein. Thus, the presence of GABA in neurons and terminals is a good initial indication that the cells in question use GABA as a neurotransmitter. The chemical structure of GABA is shown in Figure 6.6.

GABA works to inhibit neural signaling. This process of inhibition must be balanced, however. Excessive inhibition of neural signaling can lead to substantial health issues including seizures. In addition to inhibiting neural signaling, GABA also plays an important role in brain development. Research indicates that GABA helps lay down important brain circuits in early development. Studies have found a link between the levels of GABA in the brain and whether or not learning is successful. This has led to GABA being given the nickname "the learning chemical."

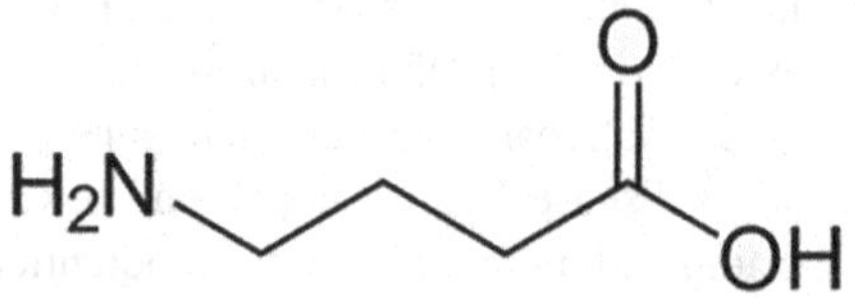

FIGURE 6.6   GABA structure.

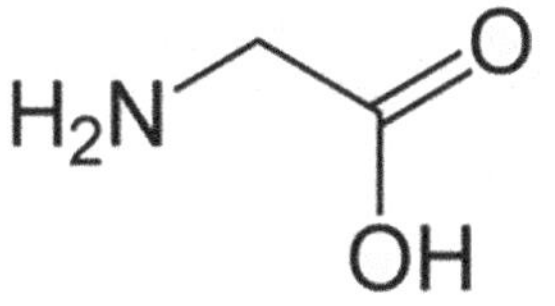

FIGURE 6.7 Glycine structure.

GLYCINE

Glycine is an amino acid that has other functions in the human body, but can also be an inhibitory neurotransmitter, particularly in the brainstem and spinal cord, as well as the retina. When glycine reaches the target cell, and that target's glycine receptors are activated, chloride enters the neurons, thus causing an inhibitory postsynaptic potential (IPSP). The chemical structure of neutral glycine is shown in Figure 6.7.

The following quote from the American College of Neuropsychopharmacology[6] might help elucidate glycine's role for you:

> Glycine is the major inhibitory neurotransmitter in the brainstem and spinal cord, where it participates in a variety of motor and sensory functions. Glycine is also present in the forebrain, where it has recently been shown to function as a coagonist at the N-methyl-D-aspartate (NMDA) subtype of glutamate receptor. In the latter, context glycine promotes the actions of glutamate, the major excitatory neurotransmitter (for a discussion of glycine's role as a coagonist of the NMDA receptor, see Excitatory Amino Acid Neurotransmission). Thus, glycine subserves both inhibitory and excitatory functions within the CNS.
>
> Glycine is formed from serine by the enzyme serine hydroxymethyltransferase (SHMT). Glycine, like GABA, is released from nerve endings in a Ca2+-dependent fashion. The actions of glycine are terminated primarily by reuptake via Na+/Cl--dependent, high-affinity glycine transporters. The specific uptake of glycine has been demonstrated in the brainstem and spinal cord in regions where there are also high densities of inhibitory glycine receptors.

High concentrations of glycine in the spinal cord were first noted in 1965.[7] Experiments showed that applying glycine to spinal neurons caused the action potential to be fired. Glycine is released from the Renshaw cell as an inhibitory neurotransmitter in the anterior horn of the spinal cord.[8]

DOPAMINE

The name dopamine is derived from a contraction if its chemical name: 3,4-dihydroxyphenethylamine. Dopamine is synthesized from its precursor molecule L-DOPA. Dopamine plays an important role in reward-motivated behavior via one of the brain's distinct dopamine pathways. There are four major dopaminergic pathways:

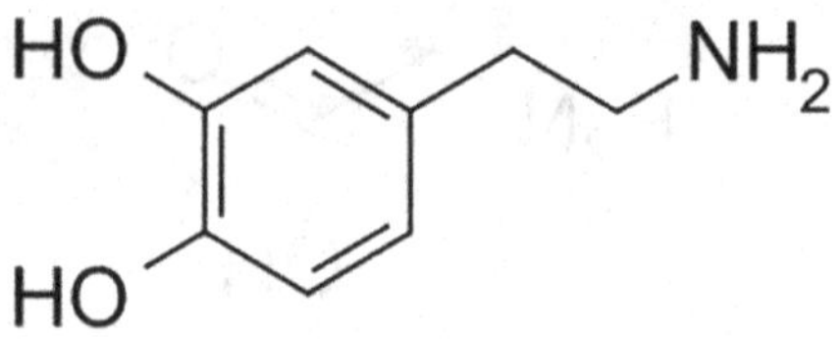

FIGURE 6.8   Dopamine structure.

1. The mesolimbic pathway: This pathway is the one related to reward functions. It connects the ventral tegmental area of the midbrain to the ventral striatum of the basal ganglia in the forebrain.
2. The mesocortical pathway: This pathway connects the ventral tegmentum to the prefrontal cortex. This pathway is thought to be related to emotional response, cognitive control, and motivation.
3. The nigrostriatal pathway: This pathway connects the substantia nigra pars compacta in the midbrain to the dorsal striatum in the forebrain. This pathway is involved in movement.
4. The tuberoinfundibular pathway. This pathway connects the arcuate nucleus in the tuberal region hypothalamus to the median eminence of the hypothalamus.

The first thing that should be clear from reviewing the dopamine pathways is that the neurotransmitter dopamine can be involved in quite different functions depending on which neurological pathway it is being used in. The structure of dopamine is shown in Figure 6.8.

The following quote regarding dopamine[9] should help you to understand its function:

> A part of the brain called the basal ganglia regulates movement. Basal ganglia in turn depend on a certain amount of dopamine to function at peak efficiency. The action of dopamine occurs via dopamine receptors, D1–5.
>
> Dopamine reduces the influence of the indirect pathway, and increases the actions of the direct pathway within the basal ganglia. When there is a deficiency in dopamine in the brain, movements may become delayed and uncoordinated. On the flip side, if there is an excess of dopamine, the brain causes the body to make unnecessary movements, such as repetitive tics.

PEPTIDE NEUROTRANSMITTERS

A peptide is a short chain of amino acids, normally 2 to 50 long. When chains are longer they are called polypeptides. There are many different types of peptides. Some are hormones that also act as neurotransmitters.[10] The first neuropeptide was discovered by Ulf von Euler and John Gaddum in 1931 and is named Substance P.[11] The large number of neuropeptide transmitters have been grouped into five general categories: the brain/gut peptides, opioid peptides, pituitary peptides, hypothalamic releasing hormones, and a final category that simply includes all other peptides not otherwise classified.[12]

The biological activity of the peptide neurotransmitters depends on their particular amino acid sequence. Propeptide precursors are typically larger than their active peptide products and can give rise to more than one species of neuropeptide. Due to this, the release of multiple neuroactive peptides from a single vesicle can stimulate complex postsynaptic responses. Some peptide transmitters have been associated in modulating emotions. Other peptide neurotransmitters are involved in the perception of pain. Still other neuropeptide transmitters, such as melanocyte-stimulating hormone, adrenocorticotropin, and β-endorphin, regulate responses to stress.

A particularly important category of peptide neurotransmitters is that of the opioids. These peptides bind to the same postsynaptic receptors activated by opium, thus their name. The active ingredients in opium include a variety of plant alkaloids, predominantly morphine. First opium, and later morphine have both been used as analgesics. Synthetic opiates such as meperidine are also used as analgesics.

The opioid peptides were discovered in the 1970s as part of a search for endogenous compounds that imitated the actions of morphine. The goal was to find other compounds that could be used as analgesics. The endogenous ligands of the opioid receptors have now been identified as a group containing more than 20 opioid peptides that fall into three classes: the endorphins, the enkephalins, and the dynorphins.

Opioid peptides are broadly dispersed throughout the brain. These peptides are frequently co-localized with other small-molecule neurotransmitters including GABA and 5-HT. In general, these peptides tend to be depressants. Opioids are also involved in complex behaviors such as sexual attraction and aggressive/submissive behaviors. They have also been associated in psychiatric disorders including schizophrenia, although the evidence for this is debated.

EPINEPHRINE AND NOREPINEPHRINE

These were formally called adrenaline and noradrenaline. Epinephrine is a hormone that is involved in regulating many different functions including respiration. Epinephrine is frequently used to treat anaphylaxis reactions (allergic reactions) and cardiac arrest. Epinephrine and norepinephrine work together to prepare the body for fight-or-flight responses to high-stress situation.

AGONISTS AND ANTAGONISTS

Neuroactive chemicals frequently are classified as agonists and antagonists. Agonists are chemicals that activate a particular receptor, producing a response. Antagonists function by blocking the activity of the receptor. Many psychoactive substances behave as agonists or antagonists.

As an example of the function of agonists, psychoactive drugs such as psylocibin (magic mushroom and LSD) are agonists at the serotonin receptors, more specifically the $5HT_{2A}$ and $5HT_{2c}$ receptors.

NEUROTRANSMITTER SYNTHESIS AND PACKING

While there are several different types of neurotransmitters, the synthesis process for neurotransmitters is similar for each of these. Precursor molecules in the cell body of the neuron are used to synthesize or derive specific neurotransmitters. As one example, the amino acid tryptophan is a precursor for the neurotransmitter serotonin.

Given the central role of neurotransmitters in all neurological function, it should be obvious that controlling the concentration of neurotransmitters in the synaptic cleft is important. Neurons therefore evolved mechanisms to regulate the synthesis, packaging, release, and removal of neurotransmitters in order to maintain the appropriate levels of neurotransmitters. The life cycle of neurotransmitters is divided into five phases:

1. Synthesis of neurotransmitters.
2. Packaging neurotransmitters into vesicles.
3. Fusion of vesicles resulting in neurotransmitter release.
4. Activation of postsynaptic receptors.
5. Removal of the neurotransmitter from the synaptic cleft.

Biology tends to make use of what is at hand, and neurobiology is no exception. In many instances, the product of breaking down neurotransmitters is reused for neurotransmitter synthesis. The enzymes necessary for neurotransmitter synthesis are made in the cell body of the presynaptic cell and are transported down the axon. Precursors are taken up into the terminals by specific transporters, and neurotransmitter synthesis and packaging take place within the nerve endings. After vesicle fusion and release, the neurotransmitter may be degraded/deconstructed via enzymes. The reuptake of the neurotransmitter starts another cycle of synthesis, packaging, release, and removal.

Peptide neurotransmitters, as well as the enzymes that alter their precursors, are synthesized in the cell body. Enzymes and propeptides are packaged into vesicles in the Golgi apparatus of the neuron. During axonal transport of these vesicles to the nerve terminals, the enzymes modify the propeptides to produce one or more neurotransmitter peptides. After vesicle fusion and exocytosis, the peptides diffuse away and are degraded by proteolytic enzymes.

The synthesis of small-molecule neurotransmitters occurs within presynaptic terminals. The enzymes needed for transmitter synthesis are synthesized in the neuronal cell body and transported to the nerve-terminal cytoplasm by a mechanism that is called slow axonal transport. Slow axonal transport operates at approximately at 0.5–5 millimeters a day. The precursor molecules used by these synthetic enzymes are typically taken into the nerve terminal by transporter proteins in the plasma membrane of the terminal. The enzymes produce a cytoplasmic pool of neurotransmitters that are then packed into synaptic vesicles by transport proteins in the vesicular membrane. For certain small-molecule neurotransmitters, the final synthetic steps actually occur inside the synaptic vesicles.

After neurotransmitters are synthesized, they are then stored in synaptic vesicles until needed. The vesicles vary for different transmitters. For example, some of the small-molecule neurotransmitters such as acetylcholine are stored in vesicles 40 to 60 nm in diameter. Neuropeptides are packaged into larger synaptic vesicles ranging 90 to 250 nm in diameter. These are just two examples of the variation in vesicles.

Once loaded with transmitter molecules, vesicles associate with the presynaptic membrane and fuse with it. This process of fusing with the membrane occurs in response to an influx of Ca2+. Recall early in this chapter the importance of ion channels was discussed. This is one example of how such ion channels facilitate neuronal activity. Although the process can occur at different rates, the essential mechanisms of vesicle release are comparable for all neurotransmitters. In general, small-molecule transmitters are secreted more rapidly than larger transmitters such as peptides. As one example, the secretion of ACh from motor neurons requires a mere fraction of a millisecond. In the other extreme, many neuroendocrine cells, including those in the hypothalamus, require high-frequency bursts of action potentials for many seconds. As a general rule, the rate of transmitter release tends to be faster at synapses using small-molecule transmitters and slower at synapses that use peptides.

After the neurotransmitter has been secreted into the synaptic cleft, it will then bind to specific receptors on the postsynaptic cell membrane. This, in turn, creates a postsynaptic electrical signal. After the transmitter has bound to the postsynaptic cell, the transmitter must then be quickly removed in order to enable the postsynaptic cell to begin another cycle of neurotransmitter release, binding, and signal generation. The details by which neurotransmitters are removed vary. However, removing neurotransmitters always involves a combination of diffusion with reuptake into nerve terminals or surrounding glial cells, degradation by transmitter-specific enzymes, or both degradation and reuptake. In the case of small-molecule neurotransmitters, specific transporter proteins remove the transmitters from the synaptic cleft, eventually returning them to the presynaptic terminal for reuse.

NEUROTRANSMITTERS AND PSYCHOACTIVE SUBSTANCES

Most psychoactive substances operate by affecting one or more neurotransmitters. Some psychoactive substances related to treatment of psychological disorders will be discussed in chapter 7. However, some common psychoactive substances will be briefly described in this section.

Cannabinoids

Cannabinoids are chemical compounds found in the cannabis plant, and also found in various animals. The cannabinoid that has received the most attention is tetrahydrocannabinol (THC) due its intoxicating effect. The chemical formula of THC is $C_{21}H_{30}O_2$, and there are multiple isomers of the molecule. It has been established that neurons have specific cannabinoid receptors. Cannabinoid receptor 1 is primarily found in neurons in the basal ganglia or the limbic system. There have, however, been CB1 receptors found in the cerebellum. Cannabinoid receptor 2 is found primarily in the immune system. THC works as a partial agonist at the cannabinoid receptor CB1. A partial agonist binds to a given receptor and activates it, but does not fully activate the receptor.

Opioids

Opioids include a number of substances all related to the poppy plant and opium. Opioids have a medical use as an analgesic or anesthetic. Opioids frequently are used for nonmedical purposes due to the euphoric effects these chemicals produce. All opioids function by binding to opioid receptors in the central nervous system. There are opioid receptors in the peripheral nervous system and even the gastrointestinal tract as well. Opioids tend to cross the blood-brain barrier easily, though some opioids cross the barrier more readily than others. Opioids such as heroin and morphine decrease the activity of GABA-releasing neurons. Morphine binds to specific μ-opioid receptors in the brain, and are agonists for those receptors.

Nicotine

Nicotine binds to nicotinic acetylcholine receptors, primarily in the brain but also in muscle tissue. Nicotine increases the level of several neurotransmitters, thus eliciting its psychoactive effects. Among the pharmacological activities of nicotine, it will bind to nicotinic receptors in the ventral tegmental area and cause the release of dopamine.

GLIAL CELLS

In addition to neurons, the nervous system also contains several types of *Glial cells*. Glial cells have a cell body and processes extending from that cell body, like neurons, but are usually smaller than neurons. Glial cells perform a range of supportive functions and are interspersed between the neurons. The primary types of glial cells found in the central nervous system are *Astroglia*, *Oligodendroglia*, and *Microglia*. The processes of *Astroglia* cells fill interstices between nerve cell processes and surrounding blood vessels. *Astroglia* cells play a role in the formation of the blood-brain barrier. They also supply nerve cells with nutrients and are involved in maintaining the ionic balance in the tissue. *Microglia* are engaged in the repair of tissue damage. In the central nervous system, *oligodendroglia* forms the myelin sheath around axons. In the peripheral nervous system, the myelin sheath consists of *Schwann cells*.

Astrocytes are a type of glial cell that outnumber neurons by five to one. The astrocytes contiguously tile the entire central nervous system (CNS) and perform a variety essential complex functions in the central nervous system. Astrocytes themselves can be divided into two main subtypes, protoplasmic or fibrous. The differentiation is based on the location and cellular morphology of the astrocyte. Protoplasmic astrocytes are found throughout all gray matter and have a morphology of several stem branches that give rise to multiple branching processes in a globoid distribution. Fibrous astrocytes are found throughout all white matter and exhibit a morphology of many long fiber-like processes.

SUMMARY

This chapter provided a summary of cellular neuroscience. The general structure of neurons was explored along with ion channels, electric current initiation and propagation, as well as the role of neurotransmitters. Understanding the individual neuron is critical to understanding neuroscience.

TEST YOUR KNOWLEDGE

1. *Acetylcholine*, the first neurotransmitter discovered
2. Three classes of neurotransmitter:

   a. Excitatory, Glial, Inhibitory

   b. Excitatory, Inhibitory, Modulatory

   c. Purkinje, Glial, Pyramidal

   d. Fast-Moving, Slow-Moving, Peptide

3. Purkinje neurons are _____ neurons.

   a. Excitatory

   b. Modulatory

   c. Inhibitory

   d. Fast-Moving

4. *Glutamate (GLU)* is the most excitatory neurotransmitter in the cortex.
5. A *peptide* is a short chain of amino acids, normally 2 to 50 long.
6. _____ are chemicals that activate a particular receptor, producing a response.

   a. Antagonists

   b. Peptides

   c. Purkinje

   d. Agonists

7. Slow axonal transport operates approximately at *0.5–5 millimeters* a day.
8. The primary types of glial cells found in the central nervous system are *Astroglia, Oligodendroglia and Microglia.*

NOTES

1. https://link.springer.com/article/10.1007/s00709-012-0407-5
2. www.sciencedirect.com/science/article/abs/pii/0014488666900239
3. www.ncbi.nlm.nih.gov/pmc/articles/PMC8053672/
4. www.news-medical.net/health/What-is-Acetylcholine.aspx
5. www.mayoclinic.org/diseases-conditions/serotonin-syndrome/symptoms-causes/syc-20354758

6. https://acnp.org/g4/GN401000008/Default.htm
7. Aprison, M. H. and Werman, R. 1965. The distribution of glycine in cat spinal cord and roots. *Life Sciences*, 4(21), 2075–2083.
8. www.sciencedirect.com/topics/biochemistry-genetics-and-molecular-biology/glycine
9. www.news-medical.net/health/Dopamine-Functions.aspx
10. www.ncbi.nlm.nih.gov/pmc/articles/PMC3918222/
11. www.sciencedirect.com/science/article/abs/pii/S1357272501000310
12. www.ncbi.nlm.nih.gov/books/NBK10873/

7 Neurological Disorders

SPECIFIC DISORDERS

It would be impossible for a single chapter in a book to deal meaningfully with all the neurological disorders that exist. Even an entire book devoted to just that topic would be of enormous size. What will be done in this section is several major neurological disorders will be summarized. The goal is to provide you sufficient information to understand the role that machine learning plays in diagnosing and treating neurological disorders.

ALS

Amyotrophic lateral sclerosis (ALS), often referred to as Lou Gehrig's disease, is a rapidly progressive neurological disease that attacks the neurons responsible for controlling voluntary muscles. ALS leads to progressive degradation of the patient's ability to control muscles, eventually including involuntary muscles as well as voluntary muscles, and leading to death. In ALS, both the upper motor neurons and the lower motor neurons degenerate. As the neurons degenerate, they cease to send messages to muscles. The lack of any stimulus to the muscles eventually leads to muscular atrophy.

As the disease progresses, the patient loses the ability to start and control voluntary movement. Symptoms usually first present in the extremities or in the swallowing muscles. Patients suffering from ALS lose their strength and the ability to move their extremities. As the disease continues, the muscles in the diaphragm and chest wall cease to function properly. This causes patients to lose the ability to breathe without ventilatory support. Although the disease does not usually impair a person's mind or personality, several recent studies suggest that some people with ALS may develop cognitive problems, such as with word fluency, decision-making, and memory.

The John Hopkins medical school has extensive online literature regarding ALS. The following excerpt is useful in understanding this disorder:

> Amyotrophic lateral sclerosis is a fatal type of motor neuron disease. It is characterized by progressive degeneration of nerve cells in the spinal cord and brain. It's often called *Lou Gehrig's disease*, after a famous baseball player who died from the disease. ALS it is one of the most devastating of the disorders that affects the function of nerves and muscles.
>
> ALS does not affect mental functioning or the senses (such as seeing or hearing), and it is not contagious. Currently, there is no cure for this disease.
>
> ALS most commonly affects people of any racial or ethnic group between the ages of 40 and 70, although it can occur at a younger age.

DOI: 10.1201/9781003230588-9

There are 2 main types of ALS:

Sporadic. This is the most common form of ALS in the U.S., making up 90% to 95% of all cases. These cases occur randomly, without any known cause, and there is no family history of ALS.

Familial. This form of ALS affects a small amount of people and is thought to be inherited.

ALS symptoms will often begin in the extremities then spread to other areas of the body. As the disease progresses, neurons die, and muscles weaken. Pain in the early stages of ALS is very rare and is still uncommon in later stages.

EPILEPSY

Epilepsy is actually a group of related disorders, which span a spectrum ranging from severe, life-threatening, and disabling, to ones that are much more benign. With epilepsy, the normal pattern of neuronal activity becomes disturbed. This can lead to myriad symptoms. The most well known are convulsions and seizures. However, epilepsy can also cause unexpected sensations, and emotional reactions. The epilepsies have many possible causes and there are several types of seizures. Anything that disturbs the normal pattern of neuron activity can induce seizures. Epilepsy root cause could be due to an abnormality in neurological connectivity (i.e., brain wiring), an imbalance of neurotransmitters, or some combination of these and other factors. EEG, MRI, and related tests are commonly used in the clinical diagnosis of epilepsy.

PARKINSON'S

Parkinson's disease is a chronic, progressive disease in which insufficient dopamine is produced. Dopamine, you will recall from chapter 6, is a neurotransmitter related to body movements. The degradation of dopamine levels leads to the patient having a lack of motor control. This results in the common symptoms of Parkinson's disease including tremors, lack of coordination, and loss of balance.

The motor symptoms are due to the death of neurons in the midbrain, specifically the substantia nigra region. The substantia nigra region is important in dopamine production, so loss of neurons in this region leads to a reduction in dopamine levels. The specific mechanisms of neuron cell death are not completely understood. It is known to involve the buildup of misfolded proteins into what are known as Lewy bodies. Lewy bodies are abnormal groupings of protein. The name, Lewy bodies, comes from the fact that Fritz Lewy in 1910 was the first to note unusual protein buildup in the brain.

The symptoms can include tremors, shuffling gate, rigidity, bradykinesia (slowness of movement), mood and cognitive dysfunction, as well as sensory issues such as altered sense of smell. Over the course of the illness, hallucinations occur in approximately 50% of patients. Anxiety issues are also quite common in Parkinson's patients.

Tourette's

Tourette's syndrome is a neurological disorder that causes people to make unintended sounds, words, and body movements, called tics. Both motor and vocal tics are generally repetitive, rapid and frequent. Tics occur suddenly, may last from several seconds to minutes, and have no meaning for the person. Tics need to be present for at least one year before the diagnosis of Tourette's syndrome can be confirmed.

Patients with Tourette's can sometimes suppress tics for a short while; however, most often, the person will eventually need to allow the tic to occur. Tics can be absent at certain times, such as during a particular class at school or a visit to a doctor, or at other times may last longer and be more severe, such as after trying to suppress them or when under stress. Tics may come and go over months, change from one tic to another, or disappear for no apparent reason.

Muscular Dystrophy

Muscular dystrophy is a group of approximately 30 diseases that cause progressive weakness and loss of muscle mass. In muscular dystrophy, abnormal genes (mutations) interfere with the production of proteins needed to form healthy muscle. There are many kinds of muscular dystrophy. A few common forms of muscular dystrophy are listed here:

- **Duchenne muscular dystrophy**: This is one of the most common and severe forms. This disease usually affects males in early childhood. Patients with the condition will usually only live into their 20s or 30s.
- **Becker muscular dystrophy**: This form of muscular dystrophy is closely related to Duchenne. However, Becker develops later in childhood and is less severe. This form of muscular dystrophy typically has less impact on life expectancy.
- **Myotonic dystrophy**: This form of muscular dystrophy can develop at any age. Fortunately, life expectancy isn't always affected.
- **Facioscapulohumeral muscular dystrophy**: This form of muscular dystrophy can develop in childhood or adulthood. The progression is slow, and it typically isn't life-threatening.
- **Limb-girdle muscular dystrophy**: This is actually a group of conditions that usually develop in late childhood or early adulthood. There are variations that develop slowly and others that can progress quickly and be life-threatening.
- **Oculopharyngeal muscular dystrophy**: Unlike other types of muscular dystrophy, this type does not typically develop until a person is between 50 and 60 years old. Furthermore, this type of muscular dystrophy doesn't tend to affect life expectancy.

Symptoms of the most common varieties of muscular dystrophy begin in childhood. Other types don't surface until adulthood. The symptoms of muscular dystrophy are:[1]

- Clumsiness
- Problems climbing stairs
- Trouble jumping or hopping
- Frequent tripping or falling
- Walking on their toes
- Leg pain
- Weakness in the face, shoulder, and arms
- Inability to open or close the eyes
- Large calves from fat buildup

Diagnosis is accomplished via the use of an electrocardiogram (EEG), muscle biopsy, and electromyogram. All of these diagnostic tests can be facilitated with machine learning algorithms. Muscular dystrophy is a progressive condition, meaning the symptoms worsen over time. It often begins by affecting a particular group of muscles, before affecting the muscles more widely. Some types of muscular dystrophy can eventually affect the heart, or the muscles used for breathing, thus becoming life-threatening.

Muscular dystrophy is genetic. It is caused by specific mutations in the genes responsible for the function and structure of muscles.

ENCEPHALITIS

Encephalitis is inflammation of the active tissues of the brain caused by an infection or an autoimmune response.[2] The inflammation causes the brain to swell, which can lead to headache, stiff neck, sensitivity to light, mental confusion, and seizures. Encephalitis strikes 10–15 people per 100,000 each year, with more than 250,000 patients diagnosed in the last decade alone in the U.S. The condition can affect anyone, but more often occurs in younger people.

The progression of this disease varies widely from mild flu-like symptoms to seizures and sensory issues. In some instances, encephalitis can be life threatening. Encephalitis can be caused by certain viruses, by bacteria, by parasites, or by issues with the immune system, which lead to autoimmune encephalitis. Like meningitis, diagnosis will often involve an EEG, spinal tap, and/or MRI scans.

DEPRESSION

There is a correlation between depression and hippocampal neurogenesis. The antidepressant fluoxetine has been shown to ameliorate depression and to stimulate neurogenesis.[3] The authors cited previous studies that demonstrated improved neurogenesis alleviated anxiety and depression. Their study also reported that selective serotonin reuptake inhibitors (SSRIs) such as fluoxetine serve two functions. The first is to enhance aerogeneration. The second function is to facilitate plasticity in neurons. This is not surprising, as serotonin is also a key regulator of cell division and differentiation. Serotonin also supports axon branching and dendritogenesis.

Neurogenesis in the hippocampus, in particular, has been correlated with depression.[4] This study also found that impaired neurogenesis is a potential substrate to disorders such as depression, but did not establish a causative link. Kraus, Castrén, Kasper, and Lanzenberger (2017) further reported that the improved neurogenesis only occurs with long-term treatment with SSRIs, not with acute treatment.

Odaira, Nakagawasai, Takahashi, Nemoto, Sakuma, Lin, and Tan-No[5] also reported evidence showing a correlation between neurogenesis and depression. Their review of literature discussed previous studies that had shown that long-term administrations of antidepressants, including SSRIs, had increased brain-derived neurotrophic factor expression in rodents. The authors began by performing a surgical procedure on anesthetized mice. The mice were olfactory bulbectomized (OBX). This resulted in depressive behaviors in the mice. The authors then treated the mice with AMPK activator, 5-aminoimidazole-4-carboxamide-1-β-d-ribonucleotide (AICAR) 7 to 14 days after the surgical procedure. The use of AICAR did ameliorate the depressive behaviors. Furthermore, postmortem examination showed that the AICAR administration increased phosphorylation and expression levels of proteins in the hippocampi of the OBX mice.

The data presented in these studies supports the conclusion that there is a correlation between neurogenesis and depression. Increased neurogenesis ameliorates depressive symptoms, and at least some antidepressants enhance neurogenesis. Conversely, reduced neurogenesis is associated with depressive symptoms. That is not to indicate that neurogenesis is the sole or even primary causative factor in depression. Merely that there is a correlation.

PROGRESSIVE SUPRANUCLEAR PALSY

This degenerative disease involves the deterioration and death of specific areas of the brain. Symptoms include slowed and unbalanced movement, difficulty moving the eyes, and cognitive impairment. The symptoms can be mistaken for those of Alzheimer's or Parkinson's. The incidence of progressive supranuclear palsy is approximately 6 in 100,000 or 0.006% of the population. Symptoms typically present later in life, often in the ages of 60 to 70. The specific etiology of progressive supranuclear palsy is unknown, but it does not appear to be genetic, as less than 1% of those diagnosed have a family member with the disorder.

Progressive supranuclear palsy affects both neurons and glial cells. The neurons display neurofibrillary tangles, which are made of clumps of the tau protein. The tangles are, however, different than those found in Alzheimer's disease. Progressive supranuclear palsy primarily affects the basal ganglia, brain stem, cerebral cortex (particularly the frontal lobes and limbic system), and the spinal cord. Progressive supranuclear palsy is divided into the following categories:

1. Classical Richardson syndrome: A late-onset neurodegenerative disease.
2. PSP-parkinsonism (PSP-P): An atypical variate of progressive supranuclear palsy that is characterized by prominent early parkinsonism rather than falling or cognitive decline.
3. PSP-corticobasal syndrome (PSP-CBS): This is another atypical variate of progressive supranuclear palsy that is characterized by a mixture of

progressive asymmetric cortical sensory loss, alien limb, asymmetric limb rigidity, dystonia, and other symptoms.

4. PSP-C: Progressive supranuclear palsy with predominant cerebellar ataxia is characterized by dystonic rigidity of the neck and upper trunk, frequent falls, and mild cognitive impairment.

5. PSP induced by Annonaceae: This is very rare, the Annonaceae are a family of flowering plants found primarily in the tropics.

According to John Hopkins,[6] the following are early signs of progressive supranuclear palsy:

- Becoming more forgetful and crankier
- Having unusual emotional outbursts, like crying or laughing at unexpected times
- Becoming angry for no real reason
- Tremors in the hands
- Trouble controlling eye movements
- Blurred vision
- Slurred speech
- Trouble swallowing
- Dementia
- Depression
- Trouble directing your eyes where you want them to go
- Inability to control the eyelids, such as unwanted blinking or being unable to open your eyes
- Trouble holding someone's gaze

Unfortunately, there is no cure for progressive supranuclear palsy. The treatment involves management of symptoms including pharmacological agents.

ALZHEIMER'S

Alzheimer's disease is a progressive neurological disorder that causes neurons to die and for the brain to atrophy. Approximately 5.8 million people over the age of 65 in the United States have Alzheimer's disease. Symptoms may begin mildly, and progress over time. These symptoms include memory loss, disorientation, mood swings, and behavioral issues.

The progression of Alzheimer's disease is associated with amyloid plaques, neurofibrillary tangles, and losing neurological connections in the brain. Amyloid plaques are extracellular deposits of the amyloid beta protein. This primarily occurs in the gray matter of the brain. Neurofibrillary tangles are aggregations of hyper-phosphorylated tau protein. The presence of neurofibrillary tangles is the primary biological marker of Alzheimer's disease. In order to have a definitive diagnosis of Alzheimer's disease, examination of the neurological tissue is necessary, and that can only be accomplished postmortem. Generally, a probable diagnosis of Alzheimer's is given based on history of illness, cognitive testing, and medical imaging.

The specific etiology of Alzheimer's is unknown. Only approximately 1 to 2% of cases seem to be inherited. Early-onset Alzheimer's disease (i.e., in the ages 30 to 50) is typically genetic and can be attributed to mutations in one of three specific genes: amyloid-beta precursor protein (APP) and presenilin's PSEN1 and PSEN2.

The following quote from the Alzheimer's Association is helpful in understanding this disease: "Alzheimer's is the most common cause of dementia, a general term for memory loss and other cognitive abilities serious enough to interfere with daily life. Alzheimer's disease accounts for 60–80% of dementia cases."

MENINGITIS

Meningitis is caused by inflammation of the meninges. Meninges are the membranes that surround the brain and spinal cord. Most sources list two types of meningitis, each with a separate cause and outcome.

Viral meningitis is the more common of the two types. It is rarely life threatening and can be spread by coughing, sneezing, or some insects. Full recovery is expected.

Bacterial meningitis is much rarer and may be fatal. It is spread through respiratory secretions such as are found in coughing. There are several bacteria that can cause bacterial meningitis. The John Hopkins University School of Medicine[7] lists four bacteria that can lead to bacterial meningitis:

- *Neisseria meningitis* (**meningococcus**). This is a common cause of bacterial meningitis in children 2 to 18 years of age. It is spread by respiratory droplets and close contact. Meningococcal meningitis occurs most often in the first year of life, but may also occur in people who live in close quarters, such as in a college dorm.
- *Streptococcus pneumoniae* (**pneumococcus**). This is the most common and most serious form of bacterial meningitis. People with weak immune systems are most at risk.
- *Haemophilus influenzae type b.* The development of the *haemophilus influenzae* type b vaccine has greatly decreased the number of cases in the U.S. Children who do not have access to the vaccine and those in day-care centers are at higher risk of getting haemophilus meningitis.
- *Listeria monocytogenes.* This has become a more frequent cause of meningitis in neonates, pregnant women, people over the age of 60, and in people of all ages who have a weak immune system.

It should be noted that while most sources list two types of meningitis, the Centers for Disease Control list an additional four types, for a total of six.[8] Those four additional types are briefly described here:

1. Fungal meningitis can develop after a fungal infection spreads from somewhere else in the body to the brain or spinal cord.
2. Parasitic Meningitis: Various parasites can cause meningitis or can affect the brain or nervous system in other ways. Overall, parasitic meningitis is much less common than viral and bacterial meningitis.

3. Primary amebic meningoencephalitis (PAM) is a rare brain infection that is caused by Naegleria fowleri and is usually fatal. Naegleria fowleri is a free-living ameba (a single-celled living organism that is too small to be seen without a microscope). From 1962 to 2021, 154 U.S. infections have been reported to the CDC, with an average of 2–3 per year. There have only been four U.S. survivors.

4. Diseases due to pathogens that spread between people, called infectious diseases, are not the only things that can cause meningitis. This page describes some of the things that can cause meningitis that do not spread from one person to another (noninfectious).

The various types of meningitis are differentiated by their etiology, not their symptoms. In all variations of the disease, the membranes covering the brain and spinal cord are infected. According to the Mayo Clinic,[9] the following are symptoms that may indicate meningitis (either type):

- Sudden high fever.
- Stiff neck.
- Severe headache.
- Nausea or vomiting.
- Confusion or trouble concentrating.
- Seizures.
- Sleepiness or trouble waking.
- Sensitivity to light.
- No appetite or thirst.
- Skin rash in some cases, such as in meningococcal meningitis.

Meningitis is usually diagnosed by one or more of the following tests: lumbar puncture, blood testing, or computed tomography scan (CAT scan). You may recall from earlier in this chapter that these are the same diagnostic steps used to diagnose encephalitis.

STROKE

A stroke is a serious neurological issue. It occurs when the blood flow to your brain is impeded significantly. According to the John Hopkins University School of Medicine, there are two types of strokes:[10]

Ischemic stroke. This type of stroke is the most common type of stroke. It happens when a major blood vessel in the brain is blocked. It may be blocked by a blood clot. Or it may be blocked by a buildup of fatty deposit and cholesterol. This buildup is called plaque.

Hemorrhagic stroke. This occurs when a blood vessel in your brain bursts, spilling blood into nearby tissues. With a hemorrhagic stroke, pressure builds up in the nearby brain tissue. This causes even more damage and irritation.

The symptoms of a stroke are the same, regardless of whether it is ischemic or hemorrhagic. A droop in facial muscles, either symmetric or asymmetric, is a common symptom. Asymmetric muscle weakness is another symptom. Difficulty in speaking is a common symptom.

MULTIPLE SCLEROSIS

This is a disease of the central nervous system that is believed to be an autoimmune disorder. The essence of multiple sclerosis is the degradation of the myelin sheath around nerve fibers. This impedes electrical conduction down the axon and thus communication with adjacent neurons and muscles. Initial symptoms can be visual (blurred or double vision), muscular (weakness in extremities, difficulty walking), fatigue, and spasms. There are four variations/types of multiple sclerosis:

Clinically isolated syndrome (CIS)
Relapsing-remitting MS (RRMS)
Primary progressive MS (PPMS)
Secondary progressive MS (SPMS)

Clinically isolated syndrome is, as the name suggests, an isolated incident. The incident appears like an MS attack, but does not fulfill all the requirements for a diagnosis of multiple sclerosis. Relapsing-remitting MS is characterized by relapses separated by periods of remission that can last months or even years. Approximately 80% of patients with multiple sclerosis at least initially have relapsing remitting type. Primary progressive is characterized by a progression of symptoms with little or no remissions. Secondary progressive MS is simply progressive MS that occurs after initially being relapsing-remitting MS.

TUMORS

In the United States, approximately 80,000 tumors are diagnosed each year, with approximately 32% being malignant/cancerous.[11] There are numerous types of brain tumors. In fact, estimates are that there are over 120 types of brain tumor.[12] A few will be described in this section. Full lists of all types of brain tumors can be found online.[13]

Astrocytomoa is cancer that can form in the brain or spinal cord. This cancer begins in astrocyte cells. Astrocytomas are classified in four grades. Grade 1—pilocytic astrocytomas are slow-growing and have well-defined borders. This is approximately 2% of all brain tumors. Grade 2—low-grade astrocytoma is also slow-growing, and rarely spreads to other parts of the central nervous system. However, unlike grade 1, grade 2 borders are not well defined. Grade 3—anaplastic astrocytomas grow faster and do invade neighboring tissue. Grade 4—glioblastoma (GBM) is the most serious grade and can be composed of multiple cell types. Median survival rate is eight months.

Chordoma is a tumor that occurs at the sacrum, near the lower end of the spine, or at the base of the skull. This type of cancer can metastasize. Central nervous system lymphoma is very aggressive and typically involves multiple tumors across the central nervous system. Medulla blastomas are usually located in the cerebellum or near the brain stem, but can spread to the spinal cord.

NEUROLOGICAL DISORDERS AND MACHINE LEARNING

Hopefully, after reading this chapter you now have a better understanding of neurological disorders, and the impact these have on patients, it is time to explore how machine learning can be utilized in this area. One of the key issues in neurological disorders is diagnosis. Diagnosis often includes imaging. This can be positron-emission tomography (PET), functional magnetic resonance imaging (fMRI), and even X-rays. However, reading these images is a nontrivial task.

Appropriate reading of medical imaging is generally left to specialist physicians. These physicians, referred to as radiologists, specialize in reading images from a variety of scans. However, being humans, there is some margin of error. There can be both false positives and false negatives. Mistakes in diagnosing neurological disorders can have substantial deleterious impacts on patients. A failure to timely diagnose a disorder can dramatically reduce the probability of the patient's treatment protocol having a positive outcome. This is an area where machine learning can be of tremendous benefit.

A 2018 article in the journal Nature stated, "Advances in imaging and recording throughput are generating neuroscience data at an ever-increasing rate, necessitating efficient data analysis approaches. This is particularly evident in subdisciplines such as connectomics, as well as the analysis of behavior or neuronal activity."[14]

Diagnostic analysis of medical imaging is one of the primary applications of machine learning in neuroscience. However, there are also applications in natural language processing and clinical predictive modeling. The use of brain-computer interfaces to facilitate mobility and communication for patients with brain disorders is becoming more common. However, the function of those interfaces often relies on machine learning. A machine learning approach allows the device to "learn" from the patient's activities, and thus provide better assistance to that patient.

SUMMARY

This chapter provides a brief overview of major neurological disorders. One of the primary uses of machine learning in neuroscience is in diagnosing disorders based on machine learning algorithms analyzing diagnostic imaging. A basic understanding of these disorders will facilitate the algorithms that will be discussed in chapters 9 through 12.

TEST YOUR SKILLS

1. *ALS* leads to progressive degradation of the patient's ability to control muscles, eventually including involuntary muscles as well as voluntary muscles, and leading to death.
2. *Parkinson's disease* is a chronic, progressive disease in which insufficient dopamine is produced.

3. Which is the most common form of muscular dystrophy?

   a. Duchenne

   b. myotonic

   c. oculopharyngeal

   d. facioscapulohumeral

4. The bacteria Streptococcus pneumoniae can be associated with what neurological disorder?

   a. Epilepsy

   b. Progressive supranuclear palsy

   c. Meningitis

   d. Parkinson's

5. _____ is another atypical variate of progressive supranuclear palsy that is characterized by a mixture of progressive asymmetric cortical sensory loss, alien limb, asymmetric limb rigidity, dystonia, and other symptoms.

   a. PSP-corticobasal syndrome (PSP-CBS)

   b. Progressive supranuclear palsy

   c. PSP induced by Annonaceae

   d. PSP-parkinsonism (PSP-P)

6. _____ astrocytomas are slow-growing and have well-defined borders.

   a. Grade 1

   b. Grade 2

   c. Grade 3

   d. Grade 4

7. Approximately 80% of patients with multiple sclerosis at least initially have which type of MS?

   a. Clinically isolated syndrome (CIS)

   b. Relapsing-remitting MS (RRMS)

   c. Primary progressive MS (PPMS)

   d. Secondary progressive MS (SPMS)

8. Which disease is most associated with reduced dopamine levels?

    a. Meningitis

    b. Encephalitis

    c. Tourette's

    d. Parkinson's

NOTES

1. www.hopkinsmedicine.org/health/conditions-and-diseases/muscular-dystrophy
2. www.hopkinsmedicine.org/health/conditions-and-diseases/encephalitis
3. Micheli, L., Ceccarelli, M., D'Andrea, G. and Tirone, F. 2018. Depression and adult neurogenesis: positive effects of the antidepressant fluoxetine and of physical exercise. *Brain Research Bulletin*, 143, 181–193.
4. Kraus, C., Castrén, E., Kasper, S. and Lanzenberger, R. 2017. Serotonin and neuroplasticity—Links between molecular, functional and structural pathophysiology in depression. *Neuroscience & Biobehavioral Reviews*, 77, 317–326.
5. Odaira, T. and Nakagawasai, O., Takahashi, K., Nemoto, W., Sakuma, W., Lin, J. R. and Tan-No, K. 2019. Mechanisms underpinning AMP-activated protein kinase-related effects on behavior and hippocampal neurogenesis in an animal model of depression. *Neuropharmacology*, 150, 121–133.
6. www.hopkinsmedicine.org/health/conditions-and-diseases/progressive-supranuclear-palsy
7. www.hopkinsmedicine.org/health/conditions-and-diseases/meningitis
8. www.cdc.gov/meningitis/index.html
9. www.mayoclinic.org/diseases-conditions/meningitis/symptoms-causes/syc-20350508
10. www.hopkinsmedicine.org/health/conditions-and-diseases/stroke
11. www.nfcr.org/blog/blog7-facts-need-know-brain-tumors/?gclid=Cj0KCQiAnsqd BhCGARIsAAyjYjT1PJktNFs8BfNsx0ogS5ZhX-dKL9y97yakp48qyqdxZ52CTaU5-DEaAsnmEALw_wcB
12. www.hopkinsmedicine.org/health/conditions-and-diseases/brain-tumor
13. https://braintumor.org/brain-tumors/about-brain-tumors/brain-tumor-types/
14. Vogt, N. 2018. Machine learning in neuroscience. *Nature Methods*, 15(1), 33–33.

8 Introduction to Computational Neuroscience

INTRODUCTION

As the name suggests, computational neuroscience involves utilizing mathematical techniques to understanding neuroscience issues. Computational neuroscience has been applied to the broad spectrum of neuroscience topics including neurophysiology, neurochemistry, and cellular neurobiology. The term computational neuroscience was coined by Professor Eric L Schwartz in 1985 in Carmel, California.

This field of neuroscience is particularly relevant to the purpose of this book: machine learning for neuroscience. While it is not the case that computational neuroscience must include machine learning, the use of machine learning in computational neuroscience is growing. In fact, there is an emerging field called computational psychiatry that synthesizes knowledge from neuroscience, machine learning, psychiatry, and psychology in order to understand psychiatric disorders.

The area of single-neuron modeling has perhaps benefited the most from computational neuroscience. There exists a large body of research showing rather detailed models of single neuron behavior. Recalling the cellular neuroscience of chapter 6, it should be readily apparent that fully understanding those mechanisms is critical to understanding all of neuroscience. It should also be noted that modelling a single neuron is less complicated than attempting to model entire brain regions. That is not to indicate that single neurons are the only focus of computational neuroscience. There have been applications of computational neuroscience to understanding memory, synaptic plasticity, sensory processing, and motor control.

The goal of computational neuroscience is to explain how electrochemical signals are utilized in the brain to process information. The goal is to explain the biophysical mechanisms of computation in neurons and neural circuits. Put another way, the brain is concerned with information processing. That processing occurs via electrochemical signals. The goal of computational neuroscience is to mathematically analyze those signals in order to have a better understanding of the information processing in the brain.

NEURON MODELS

There are multiple models of neurons. Some sources simply divide the neuron models into two categories, the first category being those models that focus on electrical input

DOI: 10.1201/9781003230588-10

and output. These are often referred to as "electrical input-output voltage models" The second category are pharmacological/natural input models. These models focus on input from either natural sources or pharmacological agents. It should be readily apparent that this latter category has a number of practical, medical applications.

One way to consider these models was described in the textbook *An Introductory Course in Computational Neuroscience*:

> The ability of neurons to convey and process information depends on their electrical properties, in particular the spatial and temporal characteristics of the potential difference across the neuron's membrane—its membrane potential. Nearly all of single neuron modeling revolves around calculating the causes and effects of changes in the membrane potential.[1]

NERNST EQUATION

The electrical potential generated across the membrane at electrochemical equilibrium, the equilibrium potential, can be predicted by a simple formula called the Nernst equation. This relationship is generally expressed as shown in equation 8.1.

$$E_X = \frac{RT}{zF} \ln \frac{[X]_2}{[X]_1} \qquad \text{(eq. 8.1)}$$

In equation 8.1, the term EX is the equilibrium potential for any ion X. R is the gas constant, which is commonly used in a wide range of chemistry applications. T is the absolute temperature (measured in Kelvin). F is the Faraday constant, which represents the amount of electrical charge in one mole of a univalent ion. The z is the valence or electrical charge of the permeant ion. The brackets indicate the concentrations of ion X on each side of the membrane. The symbol ln indicates the natural logarithm of the concentration gradient. To make this calculation easier, the formula is often converted to base 10 logarithm and room temperature as shown in equation 8.2.

$$E_X = \frac{58}{z} \log \frac{[X]_2}{[X]_1} \qquad \text{(eq. 8.2)}$$

In equation 8.2, log indicates the base 10 logarithm of the concentration ratio.

The equilibrium potential is typically defined in terms of the potential difference between the reference compartment, and the other side. For example, the exterior of the cell is the conventional reference point (defined as zero potential). Therefore, when the concentration of K+ is higher inside than out, an inside-negative potential is measured across the K+-permeable neuronal membrane.

For a basic hypothetical system that has only one permeant ion species, the Nernst equation permits the electrical potential across the membrane at equilibrium to be predicted exactly. Of course, neurons don't usually have only one permeant ion species.

GOLDMAN EQUATION

As was mentioned in the previous section, typically neurons are permeable to more than one ion. In this case the resting potential can be determined from the Goldman equation, shown in equation 8.3.

$$E_\mathrm{m} = \frac{RT}{F} \ln \left(\frac{\sum_{i}^{N} P_{\mathrm{M}_i^+} \left[\mathrm{M}_i^+ \right]_\mathrm{out} + \sum_{j}^{M} P_{\mathrm{A}_j^-} \left[\mathrm{A}_j^- \right]_\mathrm{in}}{\sum_{i}^{N} P_{\mathrm{M}_i^+} \left[\mathrm{M}_i^+ \right]_\mathrm{in} + \sum_{j}^{M} P_{\mathrm{A}_j^-} \left[\mathrm{A}_j^- \right]_\mathrm{out}} \right) \qquad \text{(eq. 8.3)}$$

This may seem a bit daunting, but it is actually a rather simple formula:

- R is the ideal gas constant.
- T is the temperature in kelvins.
- F is the Faraday's constant.
- E_m is the membrane potential measure in volts.
- P_ion is the permeability for that ion measured in meters per second.
- $[\mathrm{ion}]_\mathrm{out}$ is the extracellular concentration of that ion measured in moles per cubic meter.
- $[\mathrm{ion}]_\mathrm{in}$ is the intracellular concentration of that ion (in moles per cubic meter).

This equation is sometimes called the *Goldman-Hodgkin-Katz voltage equation*. Both the Nernst equation and the Goldman equation are two common equations for determining the potential across a membrane. Many of the elements in the equation should be familiar to readers within introductory chemistry and physics courses. For example, the ideal gas constant and the Faraday constant should be familiar.

ELECTRICAL INPUT-OUTPUT VOLTAGE MODELS

In this section we will briefly look at exemplary models in this category. There are several such models. While they have similar features and functions, they do have different purposes, advantages, and disadvantages. Most of these models concern electrical currents in neurons, either individual or in groups. Current is the basis for neurological activity, as you should recall from chapters 5 and 6.

As was described in chapter 6, the current depends on the conductance of ion channels. These ion channels can vary with the membrane potential. This actually leads to feedback loops. In these feedback loops, the membrane potential influences channel conductance, which in turn effects membrane potential. These feedback loops lead to all sorts of behaviors such as voltage spikes, and oscillations.

Hodgkin-Huxley

The Hodgkin-Huxley model describes the initiation and propagation of action potentials in neurons. This model is sometimes referred to as the conductance-based model. The model itself is a set of differential equations used to model the electrical characteristics of neurons. This model has also been applied to other electrically excitable cells such as muscle cells. The Hodgkin-Huxley model begins by treating the components of the neuron as electrical elements. The lipid bilayer is a capacitor. Voltage-gated ion channels are electrical conductors. The electrochemical gradients driving the flow of ions are voltage sources. Ion pumps are current sources. This allows one to mathematically describe the current in the neuron. This modeling provides for different formulas for various aspects of the neuron's behaviors. Equation 8.3 describes the current flowing through the lipid bilayer.

$$I_C = C_m \frac{dv_m}{dt}$$
(eq. 8.3)

This equation is showing the current flowing through the membrane (Ic), which is equal to the capacitance of the membrane (Cm) multiplied by the derivative of the membrane potential (Vm) related to time. This describes the current for the membrane. To readers with a background in basic electrical engineering or physics, this formula may seem quite familiar. This is, of course, only one formula from the Hodgkin-Huxley model. The intent of this chapter is not to provide an in-depth analysis of every computational-neuroscience technique. Rather, the intent is to introduce the reader to the field of computational neuroscience, and to explore how it relates to machine learning.

The history of the Hodgkin-Huxley model is interesting. In 1952, Alan Hodgkin and Andrew Huxley published their model, and in 1963 were given the Nobel Prize in Physiology or Medicine for their work. While this work led to tremendous scientific achievements, it began with rather unimposing work. Hodgkin and Huxley had worked together for many years. Their work focused on studying the axon of a giant squid. This led to many discoveries, such as the presence of different types of ion currents. The use of giant squids has been quite common in neuroscience research, particularly before the advent of modern imaging technology. For more details on this model, there are online resources available.[2,3]

FitzHugh-Nagumo Model

This model was introduced in the early 1960s as a simplification of the Hodgkin-Huxley model. The FitzHugh-Nagumo model (often simply referred to as FHN) is an example of a relaxation oscillator. For those readers not familiar with relaxation oscillators, in electrical engineering these are oscillators that are nonlinear and produce a non-sinusoidal wave. In the FHN model, if the external stimulus (often denoted by I_{ext}) exceeds a threshold value, then the system will exhibit an excursion in phase space, before the variables such as membrane voltage relax back to their resting values. The FHN model can be derived from Hodgkin-Huxley by first combining the variables V and m into a single variable v, then combining the variables n and h into a single variable r. There is a very clear online paper devoted to the FHM model from Science Direct.[4]

Leaky Integrate-and-Fire

Another model is the leaky integrate-and-fire model. This model was developed by Louis Lapicquie, a neuroscientist of the first half of the 20th century. This model uses the term leak to reflect the diffusions of ions through the membrane. This is related to the alternative model, aptly named the non-leaky integrate-and-fire model. The basic model for diffusion of ions through the membrane is shown in equation 8.4:

$$C_m \frac{dV_m(t)}{dt} = I(t) - \frac{V_m(t)}{R_m} \qquad \text{(eq. 8.4)}$$

Again, this formula includes a derivative with respect to time. In equation 8.4, Vm is the voltage across the cell membrane, t represents time, and Rm is the resistance of the cell membrane. I, as is typical in electronic diagrams, represents current.

Adaptive Integrate-and-Fire

The adaptive integrate-and-fire model combines elements of the leaky integration with one or more variables that are related to adaptation. One of the flaws in the previously described leaky integrate-and-fire model is it does not account for neuronal adaptation. The adaptive integrate-and-fire model does integrate that adaptation. There are primarily two formulas related to the adaptive integrate-and-fire models, shown in equations 8.5 and 8.6.

$$\tau_m \frac{du}{dt} = f(u) - R \sum_k w_k + RI(t) \qquad \text{(eq. 8.5)}$$

$$\tau_k \frac{dw_k}{dt} = a_k \left(u - u_{rest} \right) - w_k + b_k \tau_k \sum_{t^{(n)}} \delta \left(t - t^{(f)} \right) \qquad \text{(eq. 8.6)}$$

Tm is a membrane time constant, V indicates voltage, t represents time, w_k is the adaptation current number with k being an index. That k index is the same used with the t variable. E_m is the resting potential. T_f is the firing time. The Greek letter delta δ is a bit more complex, it denotes the Dirac delta function. The Dirac delta distribution is a function over real numbers. It is a tool invented by physicist Paul Dirac that is used to normalize state vectors. There are also variations of this model, such as the fractional-order leaky integrate-and-fire model.

Noisy Input Model

The noisy input model, sometimes called the diffusive noise model, relates to the spike input that a neuron receives from other neurons. These spikes are stochastic and can even be nonlinear. Equation 8.7 describes this model, including the input from other neurons.

$$\tau_m \frac{dV}{dt} = f(V) + RI(t) + RI^{noise}(t) \qquad \text{(eq. 8.7)}$$

The I(t) is the input current, and the $I^{noise}(t)$ describes uncontrolled background input. T, as with previous models, refers to time. The f(v) is perhaps the simplest part, it is just the function in question here. If you wish to delve deeper into this model, there is an excellent online resource.[5]

Hindmarsh–Rose Model

The Hindmarsh–Rose model is used to study spiking-bursting behavior of the neurons' membrane potential.[6] The mathematics for this model require a bit more mathematical maturity than some of the previous models we have examined. This model depends on differential equations. If you are not familiar with differential equations, you may feel free to skip this model. The Hindmarsh–Rose model is a system of three ordinary differential equations shown in equation 8.8.

$$\frac{dx}{dt} = y + \phi(x) - z + I$$
$$\frac{dy}{dt} = \psi(x) - y \qquad\qquad \text{(eq. 8.8)}$$
$$\frac{dz}{dt} = r\left[s\left(x - x_R\right) - z\right]$$

The variables $y(t)$ and $z(t)$ account for the transport of ions across the membrane via ion channels. For those readers wishing to delve deeper into this model, there is an excellent online demonstration at the Wolfram Demonstrations Project.[7]

Morris–Lecar Model

Catherine Morris and Harold Lecar developed this model of neurons to describe the oscillatory behavior related to Ca^{++} and K^+ ion conductance in muscle fiber. This model also requires differential equations and may not be accessible to all readers. The model is shown in equation 8.9.

$$C\frac{dV}{dt} = I - g_L\left(V - V_L\right) - g_{Ca}M_{ss}\left(V - V_{Ca}\right) - g_K N\left(V - V_K\right)$$
$$\frac{dN}{dt} = \frac{N_{ss} - N}{\tau_N} \qquad\qquad \text{(eq. 8.9)}$$

The parameters of this equation are explained here:

- I is the applied current.
- C represents membrane capacitance.
- g_L, g_{Ca}, g_K: leak, Ca^{++}, and K^+ conductances through membrane channel.
- V_L, V_{Ca}, V_K: equilibrium potential of relevant ion channels.
- V_1, V_2, V_3, V_4: tuning parameters for steady state and time constant.

As with previous models discussed in this chapter, many of the components of the formula for this model should be clear to any reader with a basic chemistry and/ or physics education. The differential equation may not be familiar to all readers. However, the purpose of this chapter is to introduce computational neuroscience to the reader. If you find you require more mathematical skills, there are numerous online resources to help refresh your knowledge of topics such as calculus,[8,9,10] including differential equations.[11,12]

GRAPH THEORY AND COMPUTATIONAL NEUROSCIENCE

Graph theory is a sub discipline of discrete mathematics that is widely used for modeling a wide range of subjects. Graph theory has been applied to numerous aspects of neuroscience.[13],[14] Graph theory has been used to study neurodegenerative diseases,[15],[16] psychiatric disorders,[17] and plays an important role in computational neuroscience.[18],[19] Graph theory also plays an important role in various machine learning tasks. Graph theory and machine learning have been used to make stock market predictions,[20] study depression,[21] and identify early-stage Parkinson's disease.[22] Graph theory is essential to artificial neural networks. These facts make graph theory an important topic in computational neuroscience. In this section, a basic introduction to graph theory will be provided. Additional resources are provided via the footnotes throughout this section.

In addition to the obvious applications of graph theory to computational neuroscience, graph theory is integral to certain types of machine learning. Some unsupervised machine learning algorithms rely on graph cliques (which will be discussed later in this section) in order to determine clustering of input data. There are even formats for creating graphs, most notably, GraphML. GraphML is Graph Modeling Language, a format that uses an XML style to generate graphs.

A finite graph G (V, E) is a pair (V, E), where V is a finite set and E is a binary relation on V (Deo, 2016). The connections between vertices are termed edges. These edges are ordered pairs and not necessarily symmetrical. Edges that connect to a specific vertex are said to be incident to that vertex. When the edge proceeds from a vertex, then it is said to be incident from that vertex. This means that the connection between vertices may be directional, in which case the edge is referred to as an arc.

When the connection is an arc, the directionality indicates the origin and endpoint of the connection. When a graph is directional, it is referred to as a digraph. If two vertices are connected via more than one edge, then this graph is considered a multiple graph or multigraph. These basic definitions provide a general description of the nature of the graph.

Related to edges that are incident to or incident from a vertex is the issue of the degree of a vertex. The degree of a vertex is simply the number of edges incident to that vertex. A vertex can connect to itself, forming a loop. Loops are also part of the degree of a given vertex. In various modeling scenarios, it is advantageous to assign some weight or cost to an arc, in which case the graph is referred to as a weighted digraph. Figure 8.1 depicts a simple digraph with a single loop.

In addition to the vertices and edges, there are incidence functions. The incidence function defines how the vertices are related. The specific objects that these vertices represent is irrelevant to the mathematics of graph theory. In fact, pure mathematics studies graphs without any specific application. Describing a graph in clearer, and more mathematical rigorous terms, is the following definition shown in equation 8.10.

$$G = (V, E, \Psi) \qquad \text{(eq. 8.10)}$$

The preceding formula simply states in mathematical terms what was described in the previous paragraph: a graph (G) is a set consisting of the vertices (V), the edges

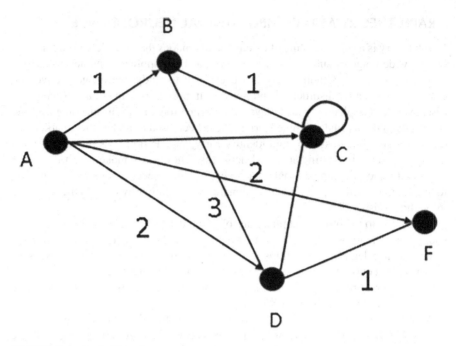

FIGURE 8.1   Basic digraph.

(E), and the incidence functions (ψ) connecting the edges to the vertices. This provides a basic description of what a graph is.

Another element of graph theory is to understand how connected the graph is. This can begin with an examination of centers of the graph. A center of a graph is a vertex with the minimal eccentricity, eccentricity being defined as the distance from the vertex to the farthest vertex in the graph. A graph may have multiple centers; therefore, the center is sometimes described as the set of vertices with minimal eccentricity.[23] Centers are interesting to modeling and analyzing neurological structure. Nuclei, described in chapter 5, are essentially graph centers.

Related to the graph center are the graph radius and diameter. The radius is the minimum of vertex eccentricities and is typically denoted as rad(G). The diameter of a graph is defined as the maximum of the vertex eccentricities and is usually denoted as diam(G). These two metrics provide valuable information about the graph. In addition to these two elementary descriptive elements of a graph, there are additional measures of central tendency that can provide insight into the system or process being graphed. The clustering coefficient is a measure of the degree to which nodes in the graph tend to cluster together.

The local clustering coefficient is often used with larger graphs. This metric quantifies how closely neighbors of a given vertex are to being a clique. The neighborhood of a vertex is generally defined as vertices immediately adjacent to the vertex in question. A clique is a subset of vertices in a graph such that every two distinct vertices in the clique are adjacent. A clique is defined as a complete graph. It should be noted that textbooks tend to focus on undirected graphs for the clustering coefficient;

there is established research applying this to directed graphs.[24] The formula for calculating the local clustering coefficient of a given vertex in a directed graph is shown in equation 8.11.

$$C_i = \frac{\left|\{e_{jk} : v_j, v_k \in N_i, e_{jk} \in E\}\right|}{k_i(k_i - 1)}$$ (eq. 8.11)

In equation 8.11, the C_i is the local clustering coefficient for a given vertex denoted by v_i. The k_i denotes the number of neighbors of a given vertex, whereas the value N_i denotes the neighborhood of the vertex. E denotes the edge set while e_{jk} is a particular edge going from vertex j to vertex k. The use of the local clustering coefficient provides insight into how connected a given vertex is. Connectivity of neurons is essential in understanding many issues in computational neuroscience.

The local clustering coefficient leads to considering the global clustering coefficient. Of particular interest in neuroscience is a variation of the global clustering coefficient named the network average clustering coefficient. This formula calculates the average of local clustering coefficients of the vertices in the graph and is shown in equation 8.12.

$$C = \frac{\Sigma_{ijk} A_{ij} A_j k A_{k_i}}{\Sigma_i k_i(k_i - 1)}$$ (eq. 8.12)

Put in simple terms, the global coefficient can be defined as the number of closed triplets divided by the number of all triplets, a triplet being three vertices that are connected. If the vertices are connected by two edges it is open, if connected by three it is closed.

In addition to measures of central tendency in a given graph, it can be useful to understand the various ways of traversing a graph. A walk is a series of vertices and edges such that the edge set connects the vertex set. The length of a walk is the number of edges. A trail is defined as a walk through the graph with no repeated edges. A path is a specialized case of a trail. A path is a trail with no repeated vertices. The length of the trails and paths give an indication of the maximal distance between vertices. The length of a path in a weighted graph is the sum of the edge-weights of the path. The distance between two vertices a and b in a weighted graph is the length of a shortest a-b path. A Eulerian trail is a trail that contains every edge of the graph G. Consider the digraph shown in Figure 8.2.

In the graph depicted in Figure 8.2, any set of steps between any two vertices is a walk. For example, the sequence C—A—B—C—B—A—F is a walk. However, a trail can have no repeated edges, thus the sequence just described would not constitute a trail. The sequence A—B—D—A—C—D—F would be a trail because no edges are repeated. But it could not be a path, because the vertices A and D are repeated. A—B—C—D—F would be a path, as no vertex or edge is repeated. These are the different methods for traversing any graph.[25] The path that is traversed via a signal is of particular importance in computational neuroscience.

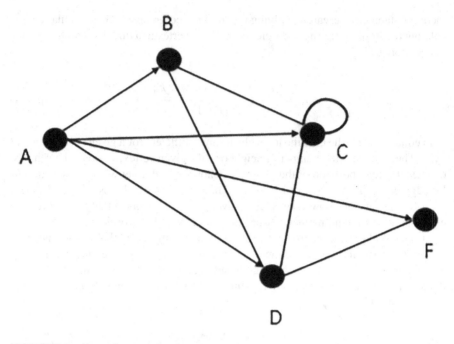

FIGURE 8.2 Exemplary graph.

ALGEBRAIC GRAPH THEORY

The basics of graph theory are primarily descriptive and not analytical. However, graph theory does provide a rich set of analytical techniques. Many of these techniques are found in the subset of graph theory known as algebraic graph theory. Algebraic graph theory takes matrix representations of graphs then applies linear algebra to those matrices.[26] This brings the power of linear algebra to bear on analyzing graphs. This allows you to combine the knowledge of chapter 1 with the information in this current chapter, to apply the combined mathematical tools to problems in neuroscience.

Graphs can also be described and analyzed using specific matrices. An incidence matrix has a row for each vertex and a column for each edge. This allows one to readily see how the edges and vertices are connected. Incidence matrices can provide insight into the connectivity in any neurological structure. The incidence matrix is calculated as shown in equation 8.13.

$$I_G[v,e] = \begin{cases} 0 \ \textit{if } v \textit{ is not an endpoint of } e \\ 1 \ \textit{if } v \textit{ is an endpoint of } e \\ 0 \ \textit{if } e \textit{ is a self} - \textit{loop at } v \end{cases} \qquad \text{(eq. 8.13)}$$

When creating the incidence matrix, vertices are listed on the left, with edges on the top. Therefore, it is first necessary to add designations for the edges to a graph. This is shown in Figure 8.3.

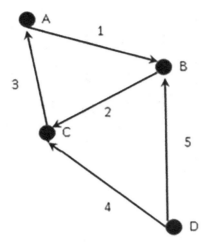

FIGURE 8.3   Adding edge designations.

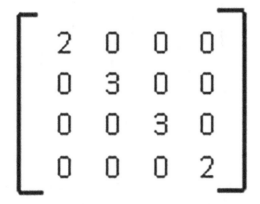

FIGURE 8.4   Incidence matrix.

The incidence matrix for the graph in Figure 8.3 is shown in Figure 8.4.

A second type of matrix used to describe a graph is the adjacency matrix. The adjacency matrix represents whether pairs of vertices are adjacent to each other. The adjacency matrix for an undirected graph is defined in equation 8.14.

$$A_G[u,v] = \begin{cases} 1 \ if \ u \ and \ v \ are \ adjacent \\ 0 \ if \ not \end{cases}$$ (eq. 8.14)

The adjacency matrix for the digraph can be calculated the same way, or it can be calculated in a similar fashion to the incidence matrix. The adjacency matrix from Figure 8.2 is shown in Figure 8.5.

In addition to the previously discussed incidence and adjacency matrix, the degree matrix is of interest. The degree matrix addresses how many connections each vertex

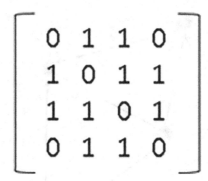

FIGURE 8.5   Adjacency matrix.

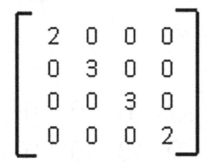

FIGURE 8.6   Degree matrix.

has. This includes self-connections/loops. The degree matrix of the graph in Figure 8.2 is provided in Figure 8.6.

It should be noted that when working with directed graphs, it is common to separately compute the in-degree matrix and the out-degree matrix. These matrices are elementary, first steps in analyzing a given graph. However, do provide information relevant to neurological interconnectivity and structure.

Another important matrix in graph theory is the Laplacian matrix. The Laplacian matrix is most often defined as the degree matrix minus the adjacency matrix and is sometimes referred to as the Kirchhoff matrix or discrete Laplacian.[27] The Laplacian matrix provides substantial information about a graph. Using Kirchhoff's theorem, it can be utilized to calculate the number of spanning trees in a graph. Thus, the Laplacian matrix for the graph in Figure 8.2 would be calculated as is shown in Figure 8.7.

There are variations of the Laplacian matrix, such as the deformed Laplacian, signless Laplacian, and symmetric normalized Laplacian. These alternatives will not be included in the methodology proposed and explored in this thesis. Specific properties of the standard Laplacian matrix will be explored. When dealing with directed graphs, one can use either the in-degree or out-degree matrix. The degree matrix demonstrates how connected specific vertices are. In neurological structures, the most connected neurons are frequently of interest.

$$\begin{bmatrix} 2 & 0 & 0 & 0 \\ 0 & 3 & 0 & 0 \\ 0 & 0 & 3 & 0 \\ 0 & 0 & 0 & 2 \end{bmatrix} \quad \begin{bmatrix} 0 & 1 & 1 & 0 \\ 1 & 0 & 1 & 1 \\ 1 & 1 & 0 & 1 \\ 0 & 1 & 1 & 0 \end{bmatrix} \quad \begin{bmatrix} 2 & -1 & -1 & 0 \\ -1 & 3 & -1 & -1 \\ -1 & -1 & 3 & -1 \\ 0 & -1 & -1 & 2 \end{bmatrix}$$

Degree Matrix - Adjacency Matrix = Laplacian Matrix

FIGURE 8.7 Laplacian matrix.

Group theory is also sometimes integrated into algebraic graph theory. This is useful in understanding graph families based on their symmetry, or lack thereof. Frucht's theorem states that all groups can be represented as an automorphism of a connected graph. This provides a means of using graphs to analyze groups. Given that neurons being analyzed can form groups, this provides a means for analyzing those groups.

Spectral Graph Theory

Spectral graph theory is a subset of algebraic graph theory. Spectral graph theory focuses on the eigenvalues, eigenvectors, and characteristic polynomials of the matrices associated with a graph.[28] Recall from chapter 1 that eigenvalues are a special set of scalars associated with a matrix that are sometimes also known as characteristic roots, characteristic values, proper values, or latent roots. As an example, consider a column vector called v. Then also consider an $n \times n$ matrix called A. Then consider some scalar λ. If it is true that $Av = \lambda v$, then v is an eigenvector of the matrix A and λ is an eigenvalue of the matrix A. Spectral graph theory uses eigenvalues to derive information about the matrix in question.

The spectrum of a matrix is its set of eigenvalues, eigenvectors, and the characteristic polynomial of the matrices associated with a graph.[29] This is normally the adjacency matrix, but in some situations the Laplacian matrix can be used. In spectral graph theory comparing the spectrum of various graph matrices provides information regarding the graphs and even changes in a graph. Spectral graph theory provides a powerful tool for analyzing graphs.

The Laplacian matrix discussed previously is often used in spectral graph theory. One of the more critical properties of a Laplacian matrix is the spectral gap of the Laplacian (Dehmer & Emmert-Streib, 2014). A spectral gap is the difference between the moduli of the two largest eigenvalues of the matrix. The spectral gap is used in conjunction with the Cheeger inequalities of a graph. Cheeger inequalities relate the eigenvalue gap with the Cheeger constant of the graph. That relationship is expressed in equation 8.15.

$$2h(G) \geq \lambda \geq \frac{h^2(G)}{2d\,max(G)} \qquad \text{(eq. 8.15)}$$

In equation 8.15, d_{max} is the maximum degree in G and λ is the spectral graph of the Laplacian matrix of the graph. The value h(G) is the Cheeger constant. The Cheeger constant is a quantitative measure of whether a graph has a bottleneck. This is sometimes called the Cheeger number and is widely used in studying networks, including both biological networks and computer or power networks. The formula for the Cheeger constant is somewhat more complex than the Cheeger inequality. Consider a graph G with a vertex set V(G) and an edge set E(G). For any set of vertices A, such that $A \subseteq V(G)$, the symbol ∂ is used to denote the set of all edges that go from a vertex in set A to a vertex outside of A. In general, the Cheeger constant is positive only if G is a connected graph. If the Cheeger constant is small and positive, that denotes some bottleneck in the graph.30

It should be noted that most definitions of both the Laplacian matrix and the Cheeger constant define these in reference to an undirected graph.[31] Most textbooks generally only discuss the Cheeger constant and the Laplacian matrix in the context of an undirected graph. This could lead to the conclusion that there is no relevance to digraphs. However, there is existing work from as early as 2005 applying both the Cheeger constant and the Laplacian to directed graphs.

The Cheeger constant is a measure of whether or not a graph has a bottleneck. Given a graph G with n vertices, the Cheeger constant is defined as shown in equation 8.16.

$$h(G) = \min_{0 < |S| \leq \frac{n}{2}} \frac{|\partial(S)|}{|S|} \qquad \text{(eq. 8.16)}$$

S is the set of nonempty sets of at most n/2 vertices and $\partial(s)$ is the edge boundary of S.

Spectral graph theory can be combined with other elements of algebraic graph theory to provide a more detailed analysis of a given graph. For example, a connected graph with diameter D will have at least D + 1 distinct values in its spectrum. This means that understanding the diameter of a graph, a rather trivial element of graph theory, provides insight into the graph's spectrum.

GRAPH SIMILARITIES

When comparing neurological structures, graph similarities can provide insight into structures that might have related functionality. Homomorphisms describe the degree of similarity between two graphs. Graph homomorphisms are also utilized extensively in a wide range of modeling applications. A graph homomorphism is a structure preserving mapping between two graphs.[32] If it is a directed graph, even the origins and tails of arcs are preserved.[33] This can be expressed with mathematical formality as shown here:

$$\text{Let } G_1 = (V_1, E_1, o_1, t_1) \text{ and } G_2 = (V_2, E_2, o_2, t_2)$$

Where V_x is the set of vertices in a graph, E_x is the set of edges, o_x is the origin of the arcs, and t_x is the tail.

$$\theta_v : V_1 -> V_2$$

$$\theta_E : E_1 -> E_2$$

Such that the origins and tails maintain their structure for all $e \in E$, is a strong homomorphism.

A strong homomorphism demonstrates a very strong degree of similarity between two graphs. However, it can be of interest to examine graph similarities that are not quite as strong. A weak homomorphism, also called a graph egamorphism, has the same edge set, but not necessarily the same origin-to-tail relationship. Put more formally, an egamorphism is a relationship for G1 and G2 such that:

if $(a,b) \in E$ and $f(a) \neq f(b)$

A graph isomorphism represents the highest degree of similarity between two graphs. An elementary definition of graph isomorphism is that if both graphs have the same number of vertices, the same number of edges, and identical degree matrices, they can be isomorphic. While these conditions are necessary for isomorphism, they are not sufficient. For two graphs V_a and V_b, there must be a function f from V_a and V_b preserves both adjacency and non-adjacency values. Essentially this requires that the two graphs have not only identical vertex sets, edge sets, and degree matrices, but that their structure is retained. Put more formally, there must exist a structure-preserving vertex bijection $f: V_a -> V_b$ in order for these two graphs to be isomorphic.

Graph similarity is also related to spectral graph theory. When comparing graphs using spectral graph theory, another important issue is that graph G is said to be determined by its spectrum if any other graph with the same spectrum as G is isomorphic to G. Put another way, if graphs G and H have the same spectrum, then they are said to be determined by that spectrum. If graphs G and H are not isomorphic but do have the same spectrum, they are said to be cospectral mates.[34]

Graph edit distance (GED) is another mechanism for comparing two graphs. This technique is related to the string edit distance used to compare two strings. The essence of graph edit distance is how many alterations need to be made to a graph G1 to make it isomorphic to graph G2. This is described mathematically in equation 8.17.

$$GED(G_1, G_2) = \min_{(e_1 \cdots e_n) \in p(G_j G_2)} \sum_{i=1}^{n} C(e_i) \qquad \text{(eq. 8.17)}$$

In equation 8.17, the e represents edges, but could be any change. The $C(e_i)$ is the change, be it an edge or vertex. The essence of the formula is simply to total the number of changes needed to alter graph g_2 such that it is isomorphic to g_1. Graph edits can be vertex or edge and involve either insertion, deletion, or substitution. This method does provide a metric of the degree of similarity between two graphs. It does not take into account incidence functions.

An area of graph theory that has not yet been covered in this chapter is spectral graph theory. This sub-domain of graph theory is focused on eigenvalues, eigenvectors, and characteristic polynomials of graph matrices. This often focuses on adjacency matrices or Laplacian matrices but can include degree matrices and incidence matrices. One important aspect of spectral graph theory for comparing systems is the area of cospectral graphs. Two graphs are considered cospectral if the adjacency matrices of the graphs have equal multisets of eigenvalues. This is sometimes termed isospectral.

Chromatic graph theory can also be useful in analyzing graphs. This aspect of graph theory is focused on the chromatic number of a graph. The chromatic number of graph G is defined as the smallest number of colors needed to color the vertex set V such that now adjacent vertices have the same color. This is often denoted as x(G). Given that loops clearly cannot be colored in this fashion, they are typically not considered in chromatic graph theory. Another part of chromatic graph theory is the chromatic polynomial. The chromatic polynomial counts the number of ways a graph can be colored, using no more k colors. Understanding the chromatic number of a graph and its chromatic polynomial gives indications as to how interconnected the graph is. For example, a high chromatic number would indicate many vertices are connected to each other, requiring more colors to prevent adjacent vertices from having the same color.

When utilizing graph theory to analyze connectivity in a neural network, a partial isomorphism is of interest. This is true for biological neural networks and for artificial neural networks. For example, the graphs of two separate neuron groups with unrelated functions are unlikely to be isomorphic. However, if a partial isomorphism is found between the two graphs, that would indicate some relation between the two neural networks. A partial isomorphism is defined as two graphs that have isomorphic subgraphs.

The degree of isomorphism is defined as the percentage to which two graphs are isomorphic. This is expressed as a percentage, rather than as an integer value. To calculate the degree of isomorphism between two graphs requires a simple formula. The number of identical vertices, edges, and incidence functions is summed. That sum divided by two, yielding the percentage of isomorphism between the two graphs. To put this in a more mathematically rigorous format, given $G1 = (V1, E2, \psi1)$ and $G2 = (V2, E2, \psi2)$, equation 8.18 demonstrates how to calculate the degree of isomorphism between the two graphs.

$$I_d = \frac{\dfrac{\left(\sum_{i=1}^{n} G_1(v) = G_2(v)\right) + \left(\sum_{i=1}^{n} G(E) = G_2(E)\right)}{n} + \dfrac{\left(\sum_{i=1}^{n} G_I(\psi) = G_2(\psi)\right)}{n}}{3}$$

(eq. 8.18)

In equation 8.18, I_d represents the degree of isomorphism. The formula is relatively simple, the number of identical vertices in each graph is divided by the total pairs of vertices being compared. The same calculation is done for edges and incidence functions. These three values are totaled and divided by three to provide a percentage of isomorphism. When two neurological structures have substantial similarity, there may also be an overlap in functionality.

Another measure of graph similarity is provided by the Randic index of a graph. The Randic index is sometimes referred to as the connectivity index. It is a measure of the degrees of the vertices in the graph. The formula is shown in equation 8.19.

$$R(G) = \sum_{(v,\omega) \in E(G)} \frac{1}{\sqrt{\delta(v)\delta(W)}} \qquad \text{(eq. 8.19)}$$

In equation 8.19, $\delta(v)$ and $\delta(W)$ denote the degrees of the vertices v and W, respectively.

When comparing graph similarities, one can utilize induced subgraphs. Rather than compare the entire graph of a network for similarities such as homomorphisms, isomorphisms, and partial isomorphisms, one can compare induced subgraphs from the network of interest.

INFORMATION THEORY AND COMPUTATIONAL NEUROSCIENCE

Essentially, the brain is an information-processing organ. Therefore, applying information theory to neuroscience,[35] particularly computational neuroscience, is a very natural application of information theory.[36] Specifically, information theory has been utilized with data mining in neuroscience.[37] It is also the case that information theory often overlaps with machine learning.[38,39,40] In this section you will be introduced to the fundamentals of information theory.

Information theory was introduced by Claude Shannon in 1948 with the publication of his article "A Mathematical Theory of Communication." Shannon was attempting to quantify information. Information theory has grown since Shannon's original paper. Shannon's seminal paper was eventually expanded into a full book. The book was entitled *The Mathematical Theory of Communication*. Shannon wrote the book with Warren Weaver and published it in 1963.[41]

In his original paper, Shannon stablished the foundational concepts of information theory. At the time, quantifying information had not yet been attempted. Shannon was particularly interested in the applications related to communicating information. The quantifying of information led to substantial changes in how information is viewed and studied.

Integrating information theory into graph theory can provide additional insights for understanding the flow between neurons and groups of neurons. Using information theory to analyze the relationship between neurons can be quite effective.

There are existing studies that have integrated information theory with algebraic graph theory. Utilizing information entropy to understand incidence functions in graphs is an existing technique. The formula for calculating Shannon entropy is shown in equation 8.20.

$$H(x) = -\sum_{i=1}^{n} p_{(x_i)} \log_2 p_{(xi)} \qquad \text{(eq. 8.20)}$$

In equation 8.20, H denotes the Shannon entropy over some variable x. The outcomes of x ($x_i, \ldots x_n$) occur a probability denoted by $p_{xi}, \ldots .p_{xn}$. While log base 2 is often assumed, it is expressly stated in equation 8.20 to avoid confusion. The information entropy provides a metric of information in a given message. When applied to the incidence function of a vertex in a given graph, it provides a metric denoting the flow of information between two vertices.

Given a graph G, the Lovász number of a graph is a metric that indicates the upper bound of the Shannon capacity of the graph.42 The Lovász number is also sometimes referred to as the Lovász theta function and is denoted by $\vartheta(G)$. The Lovász number of a graph G is defined as shown in equation 8.21.

$$\vartheta(G) = \min_{cU} \max_{i \in v} \frac{1}{\left(c^T u_i\right)^2} \qquad \text{(eq. 8.21)}$$

In equation 8.21, c denotes a unit vector in R^N. U denotes an orthonormal representation of the graph G in R^N. An orthonormal representation of G in R^N is an ordered set of unit vectors $U = (u_i \mid i \in V)$ if u_i and u_j are orthogonal when i and j are not adjacent in the graph G. There are alternative ways of depicting the Lovász-number formula. One such method is shown in equation 8.22.

$$\vartheta(G) = \max_{A \in R} \left[1 - \frac{\lambda_1(A)}{\lambda_i(A)} \right] \qquad \text{(eq. 8.22)}$$

In equation 8.22, R is the family of real matrices A(i,j) where $a_{ij} = 0$ if i and j are adjacent. The symbol λ denotes eigenvalues of A. Regardless of the form of the equation used, the result is a metric denoting the amount of information that can be transmitted via a noisy communication channel represented by the graph G.

Hartley entropy was introduced by Ralph Hartley in 1928 and is often referred to as the Hartley function. If a sample is randomly selected from a finite set A, the information revealed after the outcome is known is given by the Hartley function, shown in equation 8.23.

$$H_0(A) := \log_b |A| \qquad \text{(eq 8.23)}$$

In the Hartley function, the H0(A) is the measure of uncertainty in set A. The cardinality of A is denoted by |A|. If the base of the logarithm is 2, then the unit of uncertainty is referred to as the Shannon. However, if the natural logarithm is used instead, then the unit is the nat. When Hartley first published this function, he utilized a base-ten logarithm. Since that time, if a base ten is used, then the unit of information is called the hartley. Hartley entropy tends to be a large estimate of entropy and is thus often referred to as max-entropy.

Another measure of uncertainty in a message is the min-entropy. This measure of entropy tends to yield a small information entropy value, never greater than the standard Shannon entropy. This formula is often used to find a lower bound for information entropy. The formula is shown in equation 8.24.

$$H = \min_{1 \leq i \leq k} \left(-\log_2 p_i\right) \qquad \text{(eq. 8.24)}$$

In equation 8.24, there exists some set A with an independent discrete random variable X taken from that set and a probability that $X = x_i$ of p_i for i–1, . . .k. When the random variable X has a min-entropy (i.e., an H value) then the probability of observing any particular value for X is less than or equal to 2^{-H}.

Another measure of information entropy is the Rényi entropy. The Rényi entropy is used to generalize four separate entropy formulas: Hartley entropy, the Shannon entropy, the collision entropy, and the min-entropy, as shown in equation 8.25.

$$H\alpha\left(x\right) = \frac{1}{1-\alpha} log\left(\sum\nolimits_{i=1}^{n} p_i^{\alpha}\right)$$
(eq 8.25)

In equation 8.25, X {\displaystyle X} X denotes a discrete random variable with possible outcomes ($1, 2, ..., n$) with probabilities p_i for i = 1 to n. Unless otherwise stated, the logarithm is base 2, and the order, α, is $0 < \alpha < 1$. The aforementioned collision entropy is simply the Rényi entropy in the special case that $\alpha = 2$, and not using logarithm base 2.

All of the information theory formulas described in this section can be utilized to provide an improved understanding of a network graph. Information entropy can be implemented as the incidence function between two vertices in a network. Information entropy can also be used to weight edges or arcs. Understanding the amount and diversity of the information flowing provides an enhanced understanding of the network data flow.

COMPLEXITY AND COMPUTATIONAL NEUROSCIENCE

Clearly the brain is a very complex system. Therefore, some understanding of complexity is required to understand computational neuroscience. One of the first steps in defining complexity is to differentiate between organized and disorganized complexity. The example of molecules in a gas is a common example of disorganized complexity. The gas molecules move in a manner that is inherently unpredictable. In any system that exhibits disorganized complexity, the subsystems or components operate with a maximum of independence. This unpredictability is a cornerstone of chaos theory.

Related to complexity is chaos theory. Chaos theory is a mathematical exploration of systems that are very sensitive to early conditions. This means minor changes in the original conditions can lead to very disproportional changes in later system states. Even in a completely deterministic system, if it is sufficiently complex, it may exhibit chaotic behavior.

This is substantially different to systems that have organized complexity. Organized complexity is common in engineered systems. Disorganized complexity is more commonly found in natural systems. One might intuitively believe that disorganized complexity is more likely to lead to emergent properties and behaviors. However, that is not necessarily true. Warren Weaver wrote a 1948 paper on disorganized complexity (DC) and organized complexity (OC). His perspective was that disorganized complexity was the situation in which the interaction between parts was accidental. Organized complexity involved interactions that were intentional. Understanding the different types of complexity is relevant to understanding the framework of the current study.

EMERGENCE AND COMPUTATIONAL NEUROSCIENCE

Many of the physiological functions of the brain described in chapter 5 are emergent properties. Many researchers view human consciousness as an emergent property. This necessitates a basic understanding of emergence in order to understand computational neuroscience.

The first class is simple emergence. Some researchers would not consider simple emergence to actually be emergence. This is the system of behavior that arises in ordered systems, those that are not complex. Simple emergence is predictable behavior that can be described by the interaction of the parts. As an example, consider an airplane. None of the individual component (wings, engine, fuselage, etc.) is the airplane. However, the behavior of the airplane is completely explainable based on the constituent parts.

The second class is weak emergence. Weak emergence arises in complex systems. Weak emergence is behavior that is either desired in the system or allowed for. Weak emergence is emergence that was expected.

The third class is that of strong emergence.[43] This emergence is usually associated with the general concept of emergence. Strong emergence is a class of emergence that was not anticipated, and is often unpredictable. From a neuroscience perspective, this is emergence that is not simply a sum of the constituent parts. Consider memory as an example. This is an emergent property of many areas of the brain working in conjunction. A small, but vocal group of researchers have contended that there is no difference between strong emergence and weak emergence.

Rainey and Jamshidi[44] described a fourth class of emergence. The authors coined the term "spooky emergence" to describe this fourth category. This class of emergence cannot be foreseen or predicted in any model of the system. This class of emergence is particularly important to studying fields such as such as evolutionary biology and neuroscience.[45] That is due to the fact that "spooky emergence" can lead to novel properties and behaviors in biological systems. The current study will be most related to the latter two classes of emergence: strong and spooky. In fact, the carbon-nanotubes defects analyzed in this current study are emergent. The systems engineering body of knowledge (SEBoK) acknowledges the first three forms of emergence. However, SEBoK lacks a classification equivalent with the "spooky emergence." The concept of emergence, regardless of the type of emergence, is that complex systems can display properties and behaviors that are not readily explained by simply the sum of individual components.

With a general understanding of what emergence is and the types of emergence that can occur, it is appropriate to explore methods for analyzing and understanding emergence. Rainey and Jamshidi endorse a method called discrete-event systems. This was actually created by Bernard Zeigler for applications in biology. The discrete-event system (DEVS) is actually a method for modeling systems of systems. According to Mittal and Martín[46]:

TABLE 2.1
Levels of Systems Specifications

Level	Name	System Specification at This Level
4	Coupled systems	Systems built from component systems with a coupling recipe
3	I/O system structure	System with state space and transitions to generate the behavior
2	I/O function	Collection of I/O pairs constituting the allowed behavior partitioned according to initial state of the system
1	I/O behavior	Collection of I/O pairs constituting the allowed behavior of the system from external black box view
0	I/O frame	I/O variables and port together with values over a time base

FIGURE 8.8 DEVS levels.

The DEVS theory as proposed by Zeigler in 1976 is made up of two orthogonal concepts:

1. Levels of systems specification: Describes how systems behave.
2. System specification formalisms: Incorporates various modeling styles, such as continuous or discrete.

This process is concerned with examining the system of systems at various levels. Considering each level independently, then, in aggregate, is the approach. Mittal and Martin also provide the following hierarchy, shown in Figure 8.8.

This approach considers three hierarchical levels to each system: atomic, coupled, and associated couplings. The atomic are elements that are self-contained and could operating separately and independently. Coupled levels are clearly those items that are inextricably connected.[47] Associated couplings are additional couplings based on the primary couplings. The DEVS process is designed to examine systems of systems but is predicated on the principle that a system of systems usually leads to emergence. Thus, the DEVS process can be used to model the system and thus examine emergent behavior. However, DEVS does not focus specifically on emergence. This methodology, however, provides one mechanism to understanding emergence related to neurophysiology.

SUMMARY

This chapter has introduced computational neuroscience. This area of neuroscience is perhaps the most directly related to machine learning. Various mathematical models of neurons were introduced. Substantial emphasis was placed on both graph theory and information theory. These two mathematical tools are commonly used in both computational neuroscience and in machine learning. This chapter should provide a rudimentary knowledge of computational neuroscience.

TEST YOUR KNOWLEDGE

1. Which of the following is most related to describing the initiation and propagation of action potentials in neurons?

   a. Renyi index

   b. Hodgkin-Huxley model

   c. Noisy input model

   d. Cheeger index

2. The difference in bits between two strings X and Y is called _____.

   a. Cheeger index

   b. Hamming weight

   c. Hamming distance

   d. Renyi index

3. The amount of information that a given message or variable contains is referred to as _____

   a. Cheeger index

   b. Hamming weight

   c. Renyi index

   d. Information entropy

4. The _____ relates to the spike input that a neuron receives from other neurons.

   a. Hodgkin-Huxley model

   b. Diffusive noise model

   c. Renyi index

   d. Hamming weight

5. The set of eigenvalues, eigenvectors, and the characteristic polynomial of the matrices associated with a graph is referred to as the _____.

   a. Graph spectrum

   b. Chromatic number

   c. Renyi index

   d. Graph sum

6. When calculating the resting potential of a membrane that is permeable to multiple ions, the best formula to use is:

   a. The Goldman equation

   b. The Nernst equation

   c. The Hodgkin-Huxley Equation

   d. The Renyi Entropy

7. The *adaptive integrate-and-fire model* combines elements of the leaky integration with one or more variables that are related to adaptation.

NOTES

1. Miller, P. (2018). *An introductory course in computational neuroscience*. MIT Press.
2. https://neuronaldynamics.epfl.ch/online/Ch2.S2.html
3. www.ncbi.nlm.nih.gov/pmc/articles/PMC3424716/
4. www.sciencedirect.com/topics/medicine-and-dentistry/fitzhugh-nagumo-model
5. https://neuronaldynamics.epfl.ch/online/Ch8.S1.html
6. https://labs.ni.gsu.edu/ashilnikov/pubs/hm_all_s.pdf
7. https://demonstrations.wolfram.com/HindmarshRoseNeuronModel/
8. www.udemy.com/courses/search/?src=ukw&q=calculus
9. www.edx.org/learn/calculus
10. www.khanacademy.org/math/calculus-1
11. www.khanacademy.org/math/differential-equations
12. www.distancecalculus.com/diffeq/
13. Sporns, O. 2003. Graph theory methods for the analysis of neural connectivity patterns. In *Neuroscience databases* (pp. 171–185). Springer. https://link.springer.com/chapter/10.1007/978-1-4615-1079-6_12
14. Sporns, O. 2022. Graph theory methods: applications in brain networks. *Dialogues in Clinical Neuroscience*. www.tandfonline.com/doi/full/10.31887/DCNS.2018.20.2/osporns
15. De Haan, W., Pijnenburg, Y. A., Strijers, R. L., van der Made, Y., van der Flier, W. M., Scheltens, P. and Stam, C. J. 2009. Functional neural network analysis in frontotemporal dementia and Alzheimer's disease using EEG and graph theory. *BMC Neuroscience*, 10(1), 1–12.
16. Zhou, C., Hu, X., Hu, J., Liang, M., Yin, X., Chen, L., . . . and Wang, J. 2016. Altered brain network in amyotrophic lateral sclerosis: a resting graph theory-based network study at voxel-wise level. *Frontiers in Neuroscience*, 10, 204. www.frontiersin.org/articles/10.3389/fnins.2016.00204/full
17. Bassett, D. S., Xia, C. H. and Satterthwaite, T. D. 2018. Understanding the emergence of neuropsychiatric disorders with network neuroscience. *Biological Psychiatry: Cognitive Neuroscience and Neuroimaging*, 3(9), 742–753. www.sciencedirect.com/science/article/abs/pii/S245190221830079X
18. Marwaha, H., Sharma, A. and Sharma, V. 2022. Role of graph theory in computational neuroscience. In *Futuristic design and intelligent computational techniques in neuroscience and neuroengineering* (pp. 86–97). IGI Global. www.igi-global.com/chapter/role-of-graph-theory-in-computational-neuroscience/294593
19. de Andrade Kalil, C., de Castro, M. C. S., Silva, D. and Cortez, C. M. 2021. Applying graph theory and mathematical-computational modelling to study a neurophysiological circuit. *Open Journal of Modelling and Simulation*, 9(2), 159–171. www.scirp.org/journal/paperinformation.aspx?paperid=108822

20. Patil, P., Wu, C. S. M., Potika, K. and Orang, M. January 2020. Stock market prediction using ensemble of graph theory, machine learning and deep learning models. In *Proceedings of the 3rd International Conference on Software Engineering and Information Management* (pp. 85–92). https://dl.acm.org/doi/abs/10.1145/3378936.3378972

21. Patil, P., Wu, C. S. M., Potika, K. and Orang, M. January 2020. Stock market prediction using ensemble of graph theory, machine learning and deep learning models. In *Proceedings of the 3rd International Conference on Software Engineering and Information Management* (pp. 85–92). https://dl.acm.org/doi/abs/10.1145/3378936.3378972; www.frontiersin.org/articles/10.3389/fpsyt.2015.00021/full

22. Yan, Y., Ai, J., Liu, T. and Yan, T. October 2019. Identification of early-stage Parkinson's disease utilizing graph theory and machine learning. In *2019 12th International Congress on Image and Signal Processing, BioMedical Engineering and Informatics (CISP-BMEI)* (pp. 1–6). IEEE. https://ieeexplore.ieee.org/abstract/document/8965905

23. Hartsfield, N. and Ringel, G. 2013. *Pearls in graph theory: a comprehensive introduction.* Courier Corporation.

24. Easttom, C. 2020. On the application of algebraic graph theory to modeling network intrusions. In *IEEE 10th Annual Computing and Communication Conference, 0424–0430.*

25. Deo, N. 2017. *Graph theory with applications to engineering and computer science.* Courier Dover Publications.

26. Godsil, C. and Royle, G. F. 2013. *Algebraic graph theory.* Springer Science & Business Media.

27. Gross, J., Yellen, J. and Zhang, P. 2013. *Handbook of graph theory.* CRC Press.

28. Knauer, U. and Knauer, K. 2019. *Algebraic graph theory: morphisms, monoids and matrices.* Walter de Gruyter Press.

29. Easttom, C. and Adda, M. 2021. Application of the spectra of graphs in network forensics. In *IEEE 11th Annual Computing and Communication Conference.*

30. Chung, F. 2005. Laplacians and the Cheeger inequality for directed graphs. *Annals of Combinatorics*, 9(1), 1–19.

31. Easttom, C. 2021. A systematic framework for network forensics based on graph theory (Doctoral dissertation, University of Portsmouth). https://pure.port.ac.uk/ws/portalfiles/portal/42749878/EasttomDissertation_841367.pdf

32. Dehmer, M. and Emmert-Streib, F. (Eds.). 2014. *Quantitative graph theory: mathematical foundations and applications.* CRC Press.

33. Easttom, C. and Adda, M. 2020. The creation of network intrusion fingerprints by graph homomorphism. *WSEAS Transactions on Information Science and Applications*, 17. DOI: 10.37394/23209.2020.17.15

34. Knauer, U. and Knauer, K. 2019. *Algebraic graph theory: morphisms, monoids and matrices.* Walter de Gruyter Press.

35. Dimitrov, A. G., Lazar, A. A. and Victor, J. D. 2011. Information theory in neuroscience. *Journal of Computational Neuroscience*, 30(1), 1–5. https://link.springer.com/article/10.1007/s10827-011-0314-3

36. Linsker, R. 1990. Perceptual neural organization: some approaches based on network models and information theory. *Annual Review of Neuroscience*, 13(1), 257–281.

37. Effenberger, F. 2013. A primer on information theory with applications to neuroscience. In *Computational medicine in data mining and modeling* (pp. 135–192). Springer. https://link.springer.com/chapter/10.1007/978-1-4614-8785-2_5

38. Gao, X., Grayden, D. and McDonnell, M. 2021. Unifying information theory and machine learning in a model of electrode discrimination in cochlear implants. *PLoS One*, 16(9), e0257568. https://journals.plos.org/plosone/article?id=10.1371/journal.pone.0257568

39. Hu, B. G. 2015. Information theory and its relation to machine learning. In *Proceedings of the 2015 Chinese Intelligent Automation Conference* (pp. 1–11). Springer. https://link.springer.com/chapter/10.1007/978-3-662-46469-4_1

40. Zheng, L. and Tian, C. 2022. Information theory and machine learning. www.mdpi.com/books/pdfdownload/book/6152

41. Easttom, C. 2022. Basic information theory. In *Modern cryptography* (pp. 51–73). Springer. https://link.springer.com/chapter/10.1007/978-3-031-12304-7_3

42. Easttom, C. 2019, October. On the application of the complexity zeta function to quantify complexity in bioengineering systems. In *Proceedings of the 2019 International Conference on Artificial Intelligence and Advanced Manufacturing* (pp. 1–5). https://dl.acm.org/doi/abs/10.1145/3358331.3358375

43. Easttom, C. 2020, January. On the application of the complexity zeta function to modelling complexity and emergence in neuro-engineering. In *2020 10th Annual Computing and Communication Workshop and Conference (CCWC)* (pp. 0431–0436). IEEE.

44. Rainey, L. B. and Jamshidi, M. (Eds.). 2018. *Engineering emergence: a modeling and simulation approach*. CRC Press.

45. Easttom, C. October 2019. On the relationship of emergence and non-linear dynamics to machine learning and synthetic consciousness. In *ECIAIR 2019 European Conference on the Impact of Artificial Intelligence and Robotics* (p. 122). Academic Conferences and Publishing Limited.

46. Mittal, S., Risco-Martín, J. L. and Zeigler, B. P. 2009. DEVS/SOA: a cross-platform framework for net-centric modeling and simulation in DEVS unified process. *Simulation*, 85(7), 419–450.

47. Easttom II, W. C. C. 2020. The effects of complexity on carbon nanotube failures (Doctoral dissertation, Capitol Technology University).

Section III

Machine Learning

9 Overview of Machine Learning

INTRODUCTION

People often conflate machine learning with artificial intelligence. While the two are closely intertwined, they are not synonymous. Artificial intelligence is concerned with developing synthetic intelligence at some level. At the more basic, and more practical level, it is concerned with expert systems. At a more advanced, and more speculative level, AI is about synthetic consciousness.

Machine learning is concerned with algorithms that improve their performance over time based on input. In other words, the algorithm learns to perform a particular task better over time. This is not generalized intelligence, or even expertise. It is improving the performance of a single, specific task. While that may not sound as exciting as expert systems and artificial consciousness, it is actually something that is working today. There are many real-world applications of machine learning. Machine learning is used in facial recognition, natural language processing, analyzing medical images, and many other tasks. Our goal in this book is to apply machine learning to various issues in neuroscience. That will include medical diagnosis, brain–computer interfaces, and related applications.

Machine learning involves a range of mathematical techniques including statistics. However, it is important to keep in mind that machine learning is not merely the application of statistics. Simply applying statistics to a given problem may indeed yield results, but that is not machine learning.

BASICS OF MACHINE LEARNING

The concept of machine learning, as was stated in the beginning of this chapter, is to provide an algorithm a means to improve its performance over time. Most machine learning algorithms can be divided into one of two categories. The first is called supervised machine learning, the second is unsupervised machine learning. The names are a bit misleading. It is not about a human directly supervising every step. It is really about the outcome. In supervised machine learning, we know what the desired outcome is, and we are training the algorithm to accurately meet a specific goal.

A great application of supervised machine learning is found in a common laboratory assignment used in machine learning courses. Students are given a dataset of images of birds. The task is to train an algorithm to recognize bird species by analyzing features such as color of plumage, beak shape/size, etc. The goal is already known. The student can easily identify a robin, hawk, blue jay, etc. The task is to train the computer algorithm to do the same.

Unsupervised machine learning is often used when we don't know what is actually in the data. We want the algorithm to find specific patterns or clusters. It will still

DOI: 10.1201/9781003230588-12

require a human to analyze the results in order to divine their meaning, but the algorithm can inform us of what patterns exist.

When utilizing machine learning, there are various terms one must be familiar with. This is applicable regardless of the type of algorithm or the purpose of the machine learning project. The first such term is domain. A domain is the area in which the machine learning project is applied. For example, when applying machine learning to diagnosing neurological disorders from brain scans, the domain is neurology. One need not be an expert in the domain in order to implement a machine learning project, but having at least some knowledge of the area is essential.

Another term you will encounter frequently is model. This term has a wide range of definitions depending on the application. For example, in Microsoft Azure, a model is defined as a file that has been trained to recognize certain types of patterns.[1] Another common definition is that a machine learning model is the output of an algorithm that includes both data and the prediction algorithm.[2] Yet another definition is that a machine learning model is the mathematical representation of the output of the training process.[3] While these varied definitions may seem divergent, even contradictory, they are actually getting to the same point. Once you have completed training an algorithm, it should then be able to accomplish whatever task it has been trained for. The output of training is the machine learning model. Thus, it does indeed combine data and the prediction algorithm. And it is the output of the training process. Furthermore, in the Azure world, it is represented as a file.

Optimization is also commonly encountered in machine learning. Optimization, mathematically, is a process of minimizing some loss function. Loss functions describe discrepancy between predictions of a given model and the actual data found in the field. Optimization, in and of itself, is not machine learning. However, optimization can utilize machine learning. Furthermore, as you will encounter in later chapters, many machine learning algorithms incorporate optimization.

SUPERVISED ALGORITHMS

With supervised machine learning, you know the outcome you wish to achieve. For example, you know that you want to identify birds. Each data point that is input has features and associated labels. The goal of a supervised machine learning algorithm is to use those features and map the input to the appropriate output labels. This is done by working with training data until the algorithm achieves sufficient accuracy. The loss function in the algorithm measures the accuracy of the algorithm. All supervised machine learning algorithms share specific steps:

1. Determine the type of data in the training set and gather a training set of data.
2. Determine the input features that will be used to evaluate the input data.
3. Choose or design an algorithm for machine learning
4. Run the training set and evaluate the accuracy.

One area of research in machine learning is to determine which algorithms have the greatest accuracy with specific types of data. For example, which algorithms are best at identifying a tumor based on MRI scans of the brain.

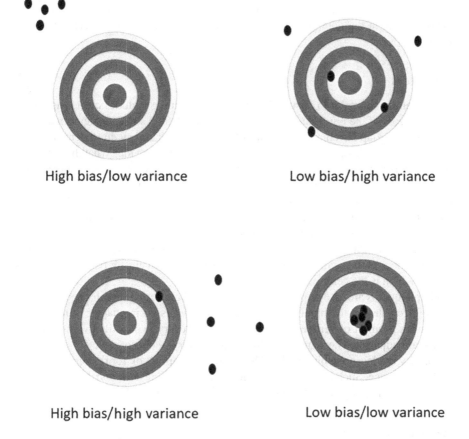

High bias/low variance   Low bias/high variance

High bias/high variance   Low bias/low variance

FIGURE 9.1   Variance and bias.

Supervised machine learning algorithms also must confront the bias–variance problem. Bias errors originate from erroneous assumptions in the algorithm. In statistics, bias refers to the difference between an expected value and the true value of a parameter being estimated. A zero bias means the algorithm is completely unbiased. Variance errors are caused by the algorithm being too sensitive to small fluctuations in the training set. Variance errors usually lead to overfitting the algorithm to the training data set. The problem is that, generally speaking, increasing bias decreases variance and vice versa. Figure 9.1 illustrates this concept.

As can be clearly seen in Figure 9.1, the ideal is to have both low bias and low variance. Any other scenario will lead to some type of error.

UNSUPERVISED ALGORITHMS

With unsupervised machine learning, one does not know that result being sought. Consider the example of supervised machine learning using identification of birds.

The classifications are known in advance, and a human being can determine if the results are correct. In unsupervised machine learning, the goal is to determine what patterns exist in the data, without a prior determination of what the classifications might be.

CLUSTERING

One common approach to unsupervised machine learning includes clustering methods. The goal of the algorithm is to group the input data in such a manner that items in the same group have more similarities to each other than they do to items in other groups. The human operating the algorithm may not have any idea what those groups will be before the algorithm is executed. This can be one of the most interesting applications of machine learning. Clustering can reveal patterns that the human operating the algorithm may not have even suspected.

There are several different clustering models that can be used. Connectivity models are based on distance. That means that each input set is a vector, and the algorithm determines the distance between input vectors to determine clusters. Another common approach is the centroid model. In this approach, each cluster has a single mean vector, which is the centroid for that cluster.

Graph models are also important. In graph-model clustering, graph cliques are the basis for clustering the input data. Cliques were discussed in chapter 8; recall that a clique is a subset of vertices in a graph such that every two distinct vertices in the clique are adjacent. These models utilize graph theory to cluster data.

Density models are a bit simpler. These models simply look for regions of density in the data space. Such regions are then determined to be clusters, and data is grouped by cluster. This approach is used in algorithms such as DBSCAN.

ANOMALY DETECTION

Anomaly detection methods are used in data analysis as well as machine learning. The concept is fairly simple. The algorithm seeks any outliers or anomalies in the data. An outlier or anomaly is any data point that deviates from the rest of the data enough to consider it maybe not being related. In statistics, outliers are sometimes removed from the data. However, when using anomaly-detection machine learning algorithms, the idea is to find the outliers.

Local outlier factor (LOF) is perhaps the most common algorithm for anomaly detection. This algorithm uses the concept of local density. LOF compares the local density of an object with that of its neighboring data points. If a data point has a lower density than its neighbors, then it is considered an outlier.

SPECIFIC ALGORITHMS

Certain algorithms will have an entire chapter or more devoted to them further in this book. For example, artificial neural networks have two chapters, k-means clustering has one chapter. In this section, we will briefly examine algorithms that do not receive such detailed treatment later in this book.

K-NEAREST NEIGHBOR

The k-nearest neighbors' algorithm (KNN) is a nonparametric method used on both regression and classification. For the purposes of malware development, it would be most useful as a classification method for improving target acquisition. When applying k-NN classification, the output is membership in a given class. Therefore, classes are predetermined. The simplest model would be target and non-target. However, that flat taxonomy can be expanded to include classifications of likely target and likely non-target. The k-nearest neighbor algorithm is essentially determining the K most similar instances to a given "unseen" observation. Similarity being defined according to some distance metric between two data points. A common choice for the distance is the Euclidean distance as shown in equation 9.1.

$$d\left(x, x^{'}\right) = \sqrt{\left(x_1 - x_1^{'}\right)^2 + \left(x_2 - x_2^{'}\right)^2 + \ldots + \left(x_n - x_n^{'}\right)^2} \qquad \text{(eq. 9.1)}$$

The algorithm functions by iterating through the entire dataset and computing the distance between x and each training observation. Then the conditional probability for each class is estimated using the function shown in equation 9.2.

$$P(y = j \mid X = x) = \frac{1}{K} \sum_{i \in A} I\left(y^{(i)} = j\right) \qquad \text{(eq. 9.2)}$$

NAÏVE BAYES

Naive Bayes classifiers are essentially classifiers that work on probabilities applying Bayes' theorem. These algorithms are well established and have been studied for decades. They are widely used for categorizing text. For example, spam filters utilize these algorithms. Conditional probabilities in Bayes' theorem are often represented by the formula shown here. The C are the classes being examined. This is shown in equation 9.3.

$$p\left(C_k \mid \mathbf{x}\right) = \frac{p\left(C_k\right) p\left(\mathbf{x} \mid C_k\right)}{p(\mathbf{x})} \qquad \text{(eq. 9.3)}$$

Naive Bayes is a relatively simple technique for classifying. Class labels are assigned to problem instances. These are represented as vectors of feature values. The features are independent variables in the formula. This is an appropriate modality for training weaponized malware based on selected feature sets for the malware.

There are a number of variations on Naive Bayes. Among those variations are the Gaussian Naive Bayes, which is often used when dealing with continuous data. Each class is distributed according to a Gaussian distribution. Multinomial Naive Bayes is used when certain events are generated by a multinomial. The specific selection of a particular version of the Naïve Bayes algorithm will be dependent on the operational needs and the goals of the training and modeling.

GRADIENT DESCENT

Gradient descent is an optimization algorithm used to minimize some function by iteratively moving in the direction of steepest descent as defined by the negative of the gradient. In machine learning, we use gradient descent to update the parameters of our model. Parameters refer to coefficients in linear regression and weights in neural networks.

Ultimately this algorithm was designed to find the minimum of a function. Starting at the top of the mountain, we take our first step downhill in the direction specified by the negative gradient. Next we recalculate the negative gradient (passing in the coordinates of our new point) and take another step in the direction it specifies. We continue this process iteratively until we get to the bottom of our graph, or to a point where we can no longer move downhill—a local minimum. The size of these steps is called the learning rate. With a high learning rate, one can cover more ground each step, but we risk overshooting the lowest point since the slope of the hill is constantly changing. Put more concisely, gradient descent is an optimization algorithm used to find the values of parameters of a function (f) that minimizes a cost function.

SUPPORT VECTOR MACHINES

Support vector machines (SVM), also called support vector networks, are supervised machine learning algorithms often used for classification. Each data point is viewed as an n-dimensional vector (i.e., a vector of n numbers). The SVM creates a hyperplane or set of hyperplanes that can be used for tasks such as classification. This, of course, necessitates defining a hyperplane. Hyperplanes are a concept borrowed form geometry. In geometry, a hyperplane is a subspace with a dimension that is one less than that of the ambient space. Mathematically, ambient space is the space surrounding some mathematical object along with the object itself. If you consider a three-dimensional space, then any hyperplanes would be two-dimensional.

In SVMs, the goal is to determine if there is a hyperplane that separates the pints in the input vector. This allows the classification of the data. Consider Figure 9.2.

In Figure 9.2, N1 effectively separates the data so that the distance from it to the nearest data point is maximized. N2 does not accomplish this goal. Therefore, N1 is

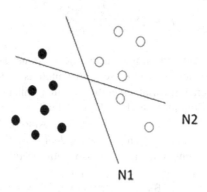

FIGURE 9.2 Hyperplanes.

used to classify the data. If an appropriate hyperplane exists, as it does in Figure 9.2, it is called the maximum-margin hyperplane. Put another way, "The objective of the support vector machine algorithm is to find a hyperplane in an N-dimensional space(N—the number of features) that distinctly classifies the data points."[4]

Support vector machines are used in image classification, recognizing handwriting, and even classifying proteins. The name stems from the fact that the support vectors are the data points that are closest to the hyperplane. These points define the separating line between classes of data points. Margins are the term for a gap between the two lines that are closest to the class points.[5] Support vector machines are supported in the scikit library,[6] which you can use in Python to create machine learning algorithms.

When working with support vector machines, another important concept is the kernel. A kernel transforms an input from the data space into the form needed. The two most common types are linear kernels and polynomial kernels. A linear kernel uses a dot product between any two input vectors (recall dot products from chapter 1). A polynomial kernel can be used to distinguish nonlinear input spaces, including curved input spaces. There are other types of kernels such as the radial basis function kernel. The radial basis function kernel is used when there is no prior knowledge regarding the input dataset.

The following code uses a support vector machine to predict the probability of an epileptic seizure. The test data can be downloaded from www.kaggle.com/code/yatindeshpande/seizure-prediction-using-svm/data

```
#import libraries
import numpy as np # linear algebra
import pandas as pd # data processing, CSV file I/O (e.g.,
pd.read _ csv)
import matplotlib.pyplot as pyplot
import seaborn as sn
import warnings
# note that this next line is importing the support vector
machine
# so, most of the work is done for you.
# SVC is the classifer
# see https://scikit-learn.org/stable/modules/generated/sklearn.
svm.SVC.html
from sklearn.svm import SVC
from sklearn.preprocessing import StandardScaler
from sklearn.model _ selection import train _ test _ split,
cross _ val _ score
#import the data. You can download the CSV from
#www.kaggle.com/code/yatindeshpande/seizure-prediction-using-
svm/data
#you will also find a similar algorithm at that location
EData = pd.read _ csv('EpilepsyData/EpilepsyData.csv')
EData = EData.drop(columns = EData.columns[0])
```

```
EData.head()
cols = EData.columns
tgt = EData.y
tgt[tgt > 1] = 0
ax = sn.countplot(tgt,label="Count")
non_seizure, seizure = tgt.value_counts()
print('The number of trials for the non-seizure class is:',
non_seizure)
print('The number of trials for the seizure class is:', seizure)
EData.isnull().sum().sum()
Y = EData.iloc[:,178].values
Y.shape
Y[Y>1]=0
Y
X = EData.iloc[:,1:178].values
X.shape
X_train, X_test, y_train, y_test = train_test_split(X, Y,
test_size = 0.3)
sc = StandardScaler()
X_train = sc.fit_transform(X_train)
X_test = sc.transform(X_test)
# if you don't provide a kernel type
# rbf is used by default.
clf = SVC(kernel='poly', degree=8)
clf.fit(X_train, y_train)
y_pred_svc = clf.predict(X_test)
acc_svc = round(clf.score(X_train, y_train) * 100, 2)
print("Accuracy is:",(str(acc_svc)+'%'))
new_input1 = [EData.iloc[6,:177]]
new_output = clf.predict(new_input1)
if new_output==[1]:
print('There is a high probability of seizure')
else:
print('There is a low probability of seizure')
```

If you execute the code as written, you should see the result shown in Figure 9.3 Note the line of code that states:

```
clf = SVC(kernel='poly', degree=8)
```

```
E:\Projects\publishing\Machine Learning For Neuroscience>python svmexample.py
The number of trials for the non-seizure class is: 9200
The number of trials for the seizure class is: 2300
Accuracy is: 96.01%
There is a high probability of seizure
```

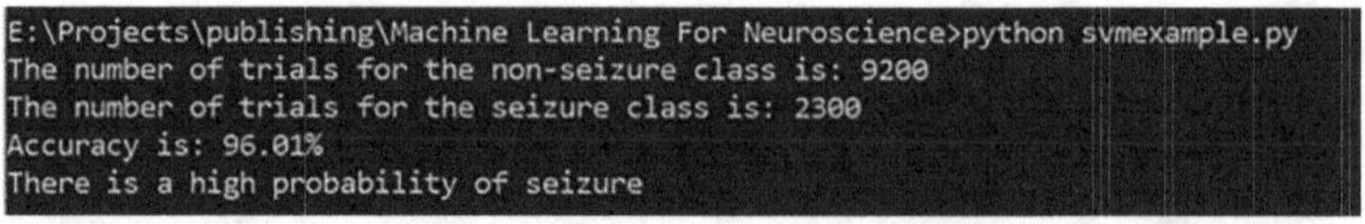

FIGURE 9.3 SVM.

This is choosing a kernel type. If you don't choose any it will use the radial basis function. To use the default, you would simply write:

```
clf = SVC()
```

If you wish to use the sigmoid kernel, try this line:

```
clf = SVC(kernel='sigmoid')
```

It is a good practice to try different kernels to compare the performance. There is a reason that there are a range of kernels to choose from. Some kernels are more appropriate for particular applications and particular datasets. Unfortunately, this part of machine learning is a bit of an art, and requires trial and error.

FEATURE EXTRACTION

Many algorithms will require you to extract the features of interest from the dataset being used. As one example, the k-nearest neighbor algorithm that we will explore in chapter 13 requires feature extraction algorithms. There are several algorithms one can use to do this; the most common are discussed in this section.

PCA

Principal component analysis (PCA) is a very widely used technique. It is particularly useful for large data sets that have a high number of features. PCA is essentially a statistical technique for reducing the dimensionality of a dataset. The principal components for a set of points in a real coordinate space are a series of p unit vectors where the ith vector is the direction of a line that best fits the data. If this description seems a bit vague to you, you should review the linear algebra in chapter 1.

PCA is not necessarily appropriate for all applications. The following quote defines three situations wherein PCA is appropriate.[7]

> Do you want to reduce the number of variables, but aren't able to identify variables to completely remove from consideration?
> Do you want to ensure your variables are independent of one another?
> Are you comfortable making your independent variables less interpretable?

Principal component analysis finds lines and planes in the K-dimensional space that approximate the data as closely as possible. Closely is defined using least squares. Least squares is a method to find the line that best fits the data. A line or plane that is the least squares approximation of a set of data points makes the variance of the coordinates on the line or plane as great as possible. Fortunately, PCA is actually built in to sklearn.decomposition. The following code will demonstrate PCA.

```
#import your modules
import numpy as np
import matplotlib.pyplot as plt
```

```
import seaborn as sns; sns.set()
from sklearn.decomposition import PCA
#create random dots to illustrate
rng = np.random.RandomState(1)
X = np.dot(rng.rand(2, 2), rng.randn(2, 100)).T
plt.scatter(X[:, 0], X[:, 1])
plt.axis('equal')
plt.show()
#you should experiment with different
#numbers of components
pca = PCA(n_components=4)
pca.fit(X)
print(pca.components_)
print(pca.explained_variance_)
def draw_vector(v0, v1, ax=None):
  ax = ax or plt.gca()
    arrowprops=dict(arrowstyle='->',
                    linewidth=2,
                    shrinkA=0, shrinkB=0)
    ax.annotate('', v1, v0, arrowprops=arrowprops)
# plot data
plt.scatter(X[:, 0], X[:, 1], alpha=0.2)
for length, vector in zip(pca.explained_variance_, pca.
components_):
v = vector * 3 * np.sqrt(length)
draw_vector(pca.mean_, pca.mean_ + v)
plt.axis('equal');
plt.show()
```

When you execute the code, you will see the following three images, Figures 9.4 to 9.6.

As you can see, sklearn.decomposition includes the PCA algorithm for you. You could implement the algorithm with even less information than is provided in this section. However, it is always best to have a general understanding of what is being done, even if you have a tool that automates the action for you.

ARTIFICIAL INTELLIGENCE

The focus of this chapter, and this book, is on machine learning. People often conflate machine learning with artificial intelligence. As was pointed out at the beginning of this chapter, that confusion is inaccurate. This section will briefly touch on the subject of artificial intelligence.

GENERAL INTELLIGENCE

Actual synthetic consciousness is not something that has been developed, or that is even close to being developed. However, more limited forms of machine intelligence

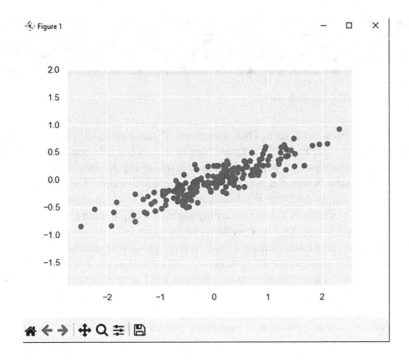

FIGURE 9.4 Data before PCA.

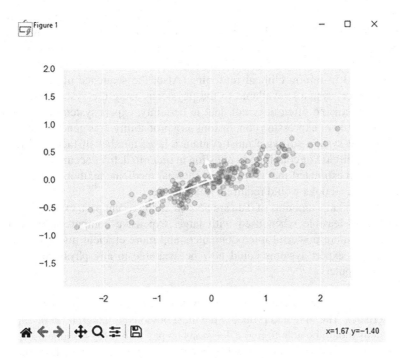

FIGURE 9.5 Data after PCA.

```
E:\Projects\publishing\Machine Learning For Neuroscience>python pca.py
[[-0.94446029 -0.32862557]
 [-0.32862557  0.94446029]]
[0.7625315 0.0184779]
```

FIGURE 9.6 PCA variance data.

can and have been developed. One such form of machine intelligence are expert systems. Expert systems utilize a knowledge base to make inferences from data. In many cases, these systems will use machine learning algorithms, such as the ones we have discussed in this chapter, as part of their functionality. The primary difference between simply applying machine learning and having an expert system is the breadth of applicability. A machine learning algorithm might learn to recognize tumors in MRI brain scans. An expert system can take that information and, along with its knowledge base, make inferences regarding prognosis and treatment.

Medical expert systems are the systems of most interest in text focused on machine learning for neuroscience. A journal article from 1987 describes medical expert systems, and this description should aid you in understanding how expert systems work:[8]

A medical expert system is a computer program that, when well-crafted, gives decision support in the form of accurate diagnostic information or, less commonly, suggests treatment or prognosis. Diagnostic, therapeutic, or prognostic advice is given after the program receives information (input) about the patient, usually via the patient's physician. Expert systems have characteristics which make them dissimilar from other kinds of medical software. One of these characteristics is that the sequence of steps used by the expert system in coming to a diagnostic or therapeutic conclusion often is designed to mimic clinical reasoning. Also, the sequence of steps is, in many expert systems, available to the physician using the system. Because clinical medicine often does not deal in certainty, expert systems may have the capability of expressing conclusions as a probability. It is generally agreed that expert system software must contain a large number of facts and rules about the disease or condition in question in order to deliver accurate answers. It has been estimated that two general internal medicine textbooks and three specialty textbooks would require 2 million rules.

Because large amounts of data are needed, in the recent past, expert systems were only feasible when used with large, expensive computers. With the advent of more powerful microcomputers and more efficient microcomputer languages, expert systems could now be available to any physician with a microcomputer.

In general, to be considered an expert system, the system must exhibit several characteristics. The first, and probably the most obvious, is a high level of expertise. The system should be able to make decisions on par with those of human experts in the field. In the case of neurology, which means an artificial system whose decisions are comparable to that of trained neurologists, the system should also be reliable and flexible. Perhaps most importantly, the system should not be prone to errors, as a human would be.

SYNTHETIC CONSCIOUSNESS

Before one can effectively address the question of synthetic consciousness, it is first instructive to examine the basis for biological consciousness. While one position of this paper is that synthetic consciousness need not necessarily be analogous to human consciousness, briefly examining human consciousness does provide a useful starting point for researching artificial consciousness. There have been a range of ideas posited regarding consciousness, ranging from the philosophical to the biological. Of particular interest for developing synthetic consciousness is the orchestrated objective reduction theory posited by Roger Penrose and Stuart Hameroff.[9] Orchestrated objective reduction (often called Orch OR) is the hypothesis that biological consciousness is an emergent property of quantum activity within the microtubules of neurons. This is a divergence from classical neurological theories that postulate biological consciousness is an emergent property of the computations performed by neurons.[10] Essentially, classical neurobiology and cognitive science see neuron activity reaching a threshold of complexity that leads to consciousness emerging. Conversely, orchestrated objective reduction sees consciousness as an emergent property derived from quantum activities within the microtubules of the neurons themselves.

Essentially, Hameroff proposed that microtubules were suitable candidates for quantum processing. Microtubules contain hydrophobic pockets that can contain delocalized electrons. Hameroff further posited that these electrons can become quantumly entangled and would form a Bose-Einstein condensate. Penrose and Hameroff's theories have been widely criticized in the neuroscientific community. Among the criticisms have been arguments that a biological system cannot avoid quantum de-coherence due to the environment of the biological system. However, advances in quantum computing have cast doubt on this criticism, as researchers have achieved quantum states for several seconds at room temperature.[11] Furthermore, evidence suggests that plants routinely use quantum-coherent electron-transport mechanisms as part of photosynthesis.[12]

Penrose and Lucas also argued that Gödel's theorem dictates that no process or algorithm can deterministically predict its outcome. Penrose further posited that this meant no algorithmic process could lead to consciousness. This was the issue that led Penrose to explore quantum behavior as the substrate for human consciousness, due to the nondeterministic nature of quantum interactions. In addition to forming a theoretical model of the basis for human consciousness, Penrose and Lucas were stating that true artificial intelligence is not possible from an algorithmic process.

Many neuroscientists disagree with Penrose, Hameroff, and Lucas. Instead, the prevailing opinion in neuroscience is that consciousness is an emergent property that is predicated on meeting a certain threshold of neurological complexity. In this view, consciousness will emerge when the neurology reaches a particular level of complexity. There is a body of evidence that supports this view. Primarily, comparative studies of neuroanatomy and physiology among diverse species show at least some association between the complexity of the brain and the level of consciousness.

Neurological complexity is more than just the number of neurons in the system. It also involves the connectivity between neurons. The complexity of the system in its entirety appears to have at least some correlation to consciousness.

Yet another neurobiological view of consciousness was posited by Nobel laureate Francis Crick. Crick posited that consciousness is inextricably associated with how the brain uses short-term memory processes to facilitate sensory input.[13] Crick focused primarily on visual input, but his work is applicable to any sensory input. His focus was on the neurological correlates of visual processing. Not merely detecting an object, but processing that sensory input.

What all of this research really means is that we do not yet have a clear understanding of the origin of human consciousness. Without that, it would be quite difficult to create synthetic consciousness. Before one can even consider research into synthetic consciousness, it is necessary to first define consciousness. Blackmore[14] states that consciousness has no generally accepted definition in science or in philosophy. This statement, while accurate, does not address the problem encountered in artificial intelligence research. Blackmore's commentary is more applicable to a complete definition of consciousness for the purposes of cognitive science and psychology. For the purposes of furthering research into artificial consciousness, it is not necessary to derive a broadly applicable, generally accepted definition of consciousness. What is required is a minimalistic definition that is operationally effective. While cognitive scientists may explore a range of definitions of consciousness,[15] for artificial intelligence research it is only necessary to select a single, operationally viable, elementary definition of consciousness. For the purposes of this paper, a simple definition of consciousness is espoused. That definition is simply self-awareness.[16] Therefore, synthetic consciousness would be operationally defined as any artificial device or software that is self-aware. This is a minimalistic definition and does not attempt to address issues regarding emotions, other related cognitive functions, or philosophical questions.

Self-awareness provides a simple definition that avoids a range of philosophical issues as well as being basic enough to be acceptable as a minimalistic operational starting point. There certainly are researchers that would add to that definition, but as a minimal definition that can be broadly accepted and operationally effective, self-awareness is an operative definition. However, even with this simple definition, there is still an issue of how to recognize and measure self-awareness. Identifying self-awareness depends on a reliable methodology. The Turing test has long been accepted as a mechanism for identifying artificial intelligence. While this has been accepted for decades, recent advances in software cast some doubt on the reliability of that testing modality. It is certainly possible now for software to be programmed to simulate a level of dialog that is difficult to differentiate from human, self-directed dialog. It should also be noted that the nature of the Turing test has an implicit assumption that consciousness must at least, at a superficial level, appear analogous to human consciousness. This is an understandable definition, given that human consciousness is the model that researchers have at hand. However, that model entails a number of complexities that are not necessary to detect consciousness. It also assumes human consciousness is the only model for consciousness. Therefore, the Turing test may not be the appropriate modality for detecting synthetic consciousness.

SUMMARY

This chapter provided a general overview of machine learning. Specific terminology and concepts were explored. The differences between supervised and unsupervised learning were explained. Several specific algorithms were also described, along with examples of those algorithms implemented in Python. This chapter forms the basis for the rest of the book, which will expand on specific machine learning algorithms and techniques.

EXERCISES

LAB 1: DETECTING PARKINSON'S

First you will download the dataset from Kaggle: www.kaggle.com/code/ vuppalaadithyasairam/feature-selection-xgboost-97-4-test-acc/data   (you will also find similar Python scripts there).

Now you will code the following script:

```
import numpy as np #
import pandas as pd # data processing, CSV file I/O
import numpy as np
from sklearn.metrics import accuracy_ score, confusion_ matrix
from sklearn.model_ selection import train_ test_ split
from xgboost import XGBClassifier
#read in data. Remember when you download you need
#to change the file name to one word with. csv not
#.xls
data=pd.read_ csv('parkinsons/Parkinssondisease.csv')
#now display the data we have
data.head()
data=data.drop(['name'],axis=1)
data.info()
corr=data.corr()
cor_ target = abs(corr["status"])
#Selecting highly correlated features
relevant_ features = cor_ target[cor_ target>0.3]
relevant_ features
data=data.drop(['MDVP:Fhi(Hz)','MDVP:Jitter(%)','MDVP:RAP','MDVP:PP
Q','Jitter:DDP','NHR','DFA'],axis=1)
data.info()
for x in data.columns:
data[x]= (data[x]-data[x].min())/(data[x].max()-data[x].min())
data.head()
y=data['status']
x=data.drop(['status'],axis=1)
```

```
X _ train,X _ test,y _ train,y _ test= train _ test _ split(x,y,test _
size=0.2,stratify=y)
svc _ model=XGBClassifier()
svc _ model.fit(X _ train,y _ train)
predictions= svc _ model. predict(X _ train)
percentage=svc _ model.score(X _ train,y _ train)
res=confusion _ matrix(y _ train,predictions)
print("Training confusion matrix")
print(res)
predictions= svc _ model. predict(X _ test)
train _ percentage=svc _ model.score(X _ train,y _ train)
test _ percentage=svc _ model.score(X _ test,y _ test)
res=confusion _ matrix(y _ test,predictions)
print("Testing confusion matrix")
print(res)
# check the accuracy on the training set
print(svc _ model.score(X _ train, y _ train))
print(svc _ model.score(X _ test, y _ test))
print(f"Train set:{len(X _ train)}")
print(f"Train Accuracy={train _ percentage*100}%")
print(f"Test set:{len(X _ test)}")
print(f"Test Accuracy={test _ percentage*100}%")
```

When you execute this, you should see output like you see here:

```
E:\Projects\publishing\Machine Learning For Neuroscience>python parkinsons.py
<class 'pandas.core.frame.DataFrame'>
RangeIndex: 195 entries, 0 to 194
Data columns (total 23 columns):
 #   Column           Non-Null Count   Dtype
---  ------           --------------   -----
 0   MDVP:Fo(Hz)      195 non-null     float64
 1   MDVP:Fhi(Hz)     195 non-null     float64
 2   MDVP:Flo(Hz)     195 non-null     float64
 3   MDVP:Jitter(%)   195 non-null     float64
 4   MDVP:Jitter(Abs) 195 non-null     float64
 5   MDVP:RAP         195 non-null     float64
 6   MDVP:PPQ         195 non-null     float64
 7   Jitter:DDP       195 non-null     float64
 8   MDVP:Shimmer     195 non-null     float64
 9   MDVP:Shimmer(dB) 195 non-null     float64
 10  Shimmer:APQ3     195 non-null     float64
 11  Shimmer:APQ5     195 non-null     float64
 12  MDVP:APQ         195 non-null     float64
 13  Shimmer:DDA      195 non-null     float64
 14  NHR              195 non-null     float64
 15  HNR              195 non-null     float64
 16  status           195 non-null     int64
 17  RPDE             195 non-null     float64
 18  DFA              195 non-null     float64
 19  spread1          195 non-null     float64
 20  spread2          195 non-null     float64
 21  D2               195 non-null     float64
 22  PPE              195 non-null     float64
dtypes: float64(22), int64(1)
memory usage: 35.2 KB
<class 'pandas.core.frame.DataFrame'>
RangeIndex: 195 entries, 0 to 194
```

Much of the output may be new to you. Some of it is specific to Parkinson's diagnosis. Some terms are defined here for you. A more comprehensive discussion of Parkinson's specific terms can be found at: www.ncbi.nlm. nih.gov/pmc/articles/PMC5434464/

D2: Correlation dimension, *a measure of the dimensionality of the space occupied by a set of random points. This comes from chaos theory.*

RPDE: Recurrence period density entropy. This is used in dynamical systems to determine periodicity.

Shimmer:DDA: Dysphonic Voice Pattern Analysis

Shimmer:APQ3: This is a three-point amplitude perturbation quotient. It is analyzing voice patterns.

NOTES

1. https://learn.microsoft.com/en-us/windows/ai/windows-ml/what-is-a-machine-learning-model
2. https://machinelearningmastery.com/difference-between-algorithm-and-model-in-machine-learning/
3. www.javatpoint.com/machine-learning-models
4. https://towardsdatascience.com/support-vector-machine-introduction-to-machine-learning-algorithms-934a444fca47
5. www.datacamp.com/tutorial/svm-classification-scikit-learn-python
6. https://scikit-learn.org/stable/modules/svm.html
7. https://towardsdatascience.com/a-one-stop-shop-for-principal-component-analysis-5582fb7e0a9c
8. https://journal.chestnet.org/article/S0012-3692(15)42851-X/fulltext
9. Hameroff, S. and Penrose, R. 2014. Consciousness in the universe: a review of the 'Orch OR' theory. *Physics of Life Reviews*, 11(1), 39–78.
10. Fingelkurts, A. A., Fingelkurts, A. A. and Neves, C. F. 2013. Consciousness as a phenomenon in the operational architectonics of brain organization: criticality and self-organization considerations. *Chaos, Solitons & Fractals*, 55, 13–31.
11. Neumann, P., Kolesov, R., Naydenov, B., Beck, J., Rempp, F., Steiner, M., . . . and Pezzagna, S. 2010. Quantum register based on coupled electron spins in a room-temperature solid. *Nature Physics*, 6(4), 249.
12. Lambert, N., Chen, Y. N., Cheng, Y. C., Li, C. M., Chen, G. Y. and Nori, F. 2013. Quantum biology. *Nature Physics*, 9(1), 10. Laureys, S., Gosseries, O. and Tononi, G. (Eds.). 2015. *The neurology of consciousness: cognitive neuroscience and neuropathology.* Academic Press.
13. Crick, F. and Koch, C. 2003. A framework for consciousness. *Nature Neuroscience*, 6(2), 119.
14. Blackmore, S. 2013. *Consciousness: an introduction.* Routledge.
15. Bermúdez, J. L. 2014. *Cognitive science: an introduction to the science of the mind.* Cambridge University Press.
16. Pope, K. (Ed.). 2013. *The stream of consciousness: scientific investigations into the flow of human experience.* Springer Science & Business Media.

10 Artificial Neural Networks

INTRODUCTION

The concept of artificial neural networks is to have an algorithm that simulates the behavior of actual mammalian neurons. That makes the study of these algorithms of particular interest in the field of neuroscience. There are many different variations of neural networks, each with its own advantages and disadvantages.

Artificial neural networks have been used for aiding systems to learn. One classic example is image recognition. Artificial neural networks can be utilized to train an algorithm to recognize images. The network is composed of nodes often called artificial neurons, thus mimicking the structure of a biological brain. Each connection can transmit a signal from one node to another node; typically this signal is a real number. The output of each node/artificial neuron is computed using a nonlinear function of the sum of the inputs.

CONCEPTS

Artificial neural networks (ANN) are designed to simulate the behavior of actual, biological networks. There are some concepts that form the foundations of ANNs. One of the most fundamental is Hebb's rule. Hebb's rule states that the changes in the strength of synaptic connections are proportional to the correlation in the firing of the two connecting neurons. Therefore, if two neurons regularly fire simultaneously, then the connection between them will be strengthened. Hebb's rule also indicates that the opposite is also true: if two neurons rarely fire simultaneously, the connection between them will weaken. This has been put more colloquially as "neurons that fire together wire together." This is true for ANNs as well. As two nodes in an artificial neural network fire together more often, the link between them becomes stronger.

Artificial neural networks consist of nodes that are connected to each other to form a network. The connections between the nodes can be directed and have weights. This is directly related to the graph theory discussed in chapter 8. In ANNs, the learning process involves the network itself changing parameters due to feedback from previous iterations. This is typically accomplished by adjusting the weightings between nodes. By changing the weightings between nodes, the likelihood of specific nodes firing in subsequent iterations is altered.

The nodes are organized into layers: the various layers work to process data. Those connections may flow in only one direction, or in both directions. The level of connectivity within an ANN also varies with different types of ANNs. As we will see in this chapter, and the next, the variation of ANNs is substantial.

ANN TERMINOLOGY

As you delve deeper into artificial neural networks, there is important terminology you should know. A basic glossary is provided here. Additional terminology will be introduced as needed in the chapter.

Activation Function is a function that determines if the inputs to a node reach the threshold to cause the node to fire to the node in the next layer. We will examine some specific activation functions later in this section.

Accuracy is a measure of the performance of classification, classification with missing inputs, and transcription; accuracy is the proportion of examples for which the model produces the correct output. Accuracy is affected by weights and biases.

Cross Entropy is a metric for estimating how well a model would generalize to new data by testing the model against one or more non-overlapping data subsets withheld from the training set.

Cross Validation is repeated use of the same data, but split differently (i.e., different training and testing sets).

Dimensionality Reduction summarizes a dataset using its common occurring patterns. Dimension reduction finds patterns in data, and uses these patterns to re-express it in a compressed form. This makes subsequent computation with the data much more efficient.

Error E is a function that computes the inaccuracies of the network as a function of the outputs y and targets t.

Error Rate is a performance metric of the classification of classification, classification with missing inputs, and transcription, the proportion of examples for which the model produces an incorrect output.

Feature measures property of an object or event with respect to a set of characteristics.

Inputs: An input vector is the data given as one input to the algorithm. This will usually be a vector (recall vectors from chapter 1). Written as x, with elements x_i, where i runs from 0 to the number of input dimensions, n.

One-Hot is a group of bits with a high bit 1 and a low bit 0. One hot encoding is used to indicate state in some systems.

Outputs: The output vector is y, with elements y_j, where j runs from 1 to the number of output dimensions, n. We can write y(x,W) to remind ourselves that the output depends on the inputs to the algorithm and the current set of weights of the network.

Overfitting occurs when comparing the complexity of hypothesis class H with the complexity of the function underlying the data, H, is too complex, and the data is not enough to constrain it.

Precision, to measure the performance of classification, is the fraction of detections reported by the model that were correct.

Stochastic Gradient Descent, also known as incremental gradient descent, allows one to approximate the gradient with a single data point instead of all available data. At each step of the gradient descent, a randomly chosen data point is used to compute the gradient direction.

Targets: The target vector t, with elements tj, where j runs from 1 to the number of output dimensions, n, are the extra data that we need for supervised learning, since they provide the "correct" answers that the algorithm is learning about.

Tensors are primary data structures in TensorFlow programs. Tensors are N-dimensional data structures, most commonly scalars, vectors, or matrices. The elements of a Tensor can hold integer, floating-point, or string values.

Training Set data is typically divided into a training set and a testing set.

Underfitting occurs when comparing the complexity of hypothesis class H with the complexity of the function underlying the data, H, is less complex than the function.

Weights wij are the weighted connections between nodes i and j. For neural networks, these weights are analogous to the synapses in the brain.

Activation Functions

Activation functions, mentioned earlier in this chapter, are key to the function of artificial neural networks. An activation function takes the input and determines if the input to the current neuron is sufficient to cause that neuron to fire the next neuron. Activation functions define the learning pattern and determine the efficiency of the algorithm. This section describes some commonly used activation functions.

Google describes activation functions as "A function (for example, ReLU or sigmoid) that takes in the weighted sum of all of the inputs from the previous layer and then generates and passes an output value (typically nonlinear) to the next layer."[1]

Rectified linear unit (or ReLU) is commonly used in Tensorflow. This function will take the input, and if it is positive, will simply output that input, with no changes. If it is not positive, then the ReLU activation function will output 0. The ReLU function is defined by the formula in equation 10.1.

$$\begin{cases} 0 \; if x \leq 0 \\ x \; if x > 0 \end{cases} \qquad \text{(eq. 10.1)}$$

ReLU is widely used but suffers from something called the dying ReLU problem. Essentially, during the training phase, some nodes/neurons cease functioning or outputting anything other than 0. Essentially, these nodes die. There are cases wherein as much as half the nodes die. One answer to that is the Leaky ReLU. This activation function is essentially a ReLU that has a parameter that determines how much the function leaks. That leakage prevents the death of nodes.

There are other variations of the ReLU function. One often used with Tensorflow is the ReLU6. This activation function has been shown to be faster than traditional ReLU. The ReLU6 function is defined by the formula in equation 10.2.

$$f(x) = \min (\max(0,x),6) \qquad \text{(eq. 10.2)}$$

The sigmoid function is also frequently seen in Tensorflow. This takes the input and compresses it into a range between 0 and 1. A result of 0 would mean the current node does not fire, whereas a 1 would mean it is fully firing. The sigmoid activation function suppresses gradients. This is sometimes called the logistic function and is defined by the formula in equation 10.3.

$$0 - (x) = \frac{1}{1 + e^x}$$
(eq. 10.3)

The sigmoid function was first published in the 1990s. The fact that it suppresses gradients was a substantial advance in activation functions.

The hyperbolic tangent (tanh) activation function is similar to sigmoid, except that it compresses the input to a range between -1 and $+1$. This activation function is defined by the formula in equation 10.4.

$$\tanh(x) = \frac{e^x - e^{-x}}{e^x + e^{-x}}$$
(eq. 10.4)

Unlike the sigmoid function, the tanh function is zero-centered. However, like the sigmoid function, it is computationally expensive due to the exponential operations in the function. You may recall from chapter 4, in the section on algorithms, the discussion of computational cost of an algorithm.

OPTIMIZATION ALGORITHMS

Optimization algorithms are used to adjust parameters in order to minimize the cost (computational cost) of the function. For each of these algorithms, this section will present a general description. The details of these algorithms are not necessary, because TensorFlow handles the mathematical implementation of the algorithm for you. This is not an exhaustive list of all optimization algorithms available in TensorFlow. This description of these key algorithms should help you understand optimization as a concept. You can view all current TensorFlow optimization algorithms at the TensorFlow website.[2]

Gradient descent is an optimization algorithm used to minimize a given function by iteratively moving in the direction of steepest descent as defined by the negative of the gradient. In machine learning, gradient descents are useful to update the parameters of the model being used. Parameters refer to coefficients in linear regression and weights in neural networks. Put in more rigorous mathematical terms, the gradient-descent algorithm is used to find the minimum of a function. Put more simply, gradient descent is an optimization algorithm used to find the values of parameters of a function that minimizes a cost function (computational cost). When using TensorFlow, the class SGD is the gradient-descent optimizer. It should be noted that gradient descent is perhaps the most common optimization algorithm used.

Another optimizer is adaptive moment estimation (ADAM). This is a variation of gradient descent. In fact, ADAM combines two different gradient-descent approaches: root mean square propagation and adaptive gradients. Rather than use the entire data set to calculate the gradient, adaptive moment estimation (ADAM) uses a randomly selected subset of the data. This creates a stochastic approximation of the gradient descent. ADAM is also a widely used optimization algorithm.

NADAM is a variation of ADAM that uses a Nesterov momentum. This, of course, necessitates a discussion of what a Nesterov momentum is. Any gradient descent can be modified with momentum. In this context, momentum is some adjustment to the gradient-descent parameter so that movement through the parameter space is averaged over multiple steps. Normally this is done by introducing velocity. The goal is that momentum will increase in those directions that lead to the most improvement. Nesterov momentum is a variation of that concept of momentum. Rather than calculate momentum with the actual positions in the search space, it calculates based on projected positions in the search space.

Adaptive gradient (AdaGrad) is actually a group of closely related algorithms. As the name suggests, it is a variation of gradient descent. A limitation of gradient descent is that it uses the same step size (learning rate) for each input variable, thus AdaGrad seeks to overcome that limitation. AdaGrad allows step size in each dimension used by the optimization algorithm to be automatically adapted based on the gradients observed for the variable.

MODELS

Models are basically files that are trained to recognize patterns. The model is trained using a set of training data. After training, the model can be used to make predictions from the data. NVIDIA, a graphics card and processor manufacturer, describes models in the following manner: "A machine learning model is an expression of an algorithm that combs through mountains of data to find patterns or make predictions. Fueled by data, machine learning (ML) models are the mathematical engines of artificial intelligence."[3] TensorFlow has an online repository of models that have already been trained for various purposes.[4]

FEEDFORWARD NEURAL NETWORKS

The name of these networks derives from the fact that outputs from nodes in one layer can only go to nodes in the next layer. There is no possibility for a cycle or loopback. The information "feeds forward," thus the name. This is among the simplest forms of a neural network. In fact, many of the concepts in the feedforward network are found in all neural networks. Thus, this particular variation bears close study.

The simplest type of feedforward network is the single-layer perceptron. The concept of a perceptron predates not only machine learning, but even digital computers. Warren McCulloch and Walter Pitts published an interesting paper in 1943, entitled "A Logical Calculus of Ideas Immanent in Nervous Activity."[5] Their goal was to simply model how neurons work. This modeling led, eventually, to the creation of artificial neurons. Each neuron takes in input, sums the input, and based on that summation, determines if it will fire or not. The model is shown in figure 10.1

Preceding nodes (X_1 through X_n) send input to the node in question. Each of these has a specific weight associated with it (W_1 through W_n). That node sums the input (thus the summation symbol) and if the sum of the weighted input meets or exceeds

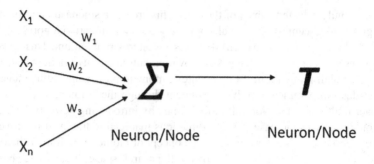

FIGURE 10.1 McCulloch–Pitts neuron.

```
1  import numpy as np
2
3  class Perceptron(object):
4
5      def __init__(self, no_of_inputs, threshold=100, learning_rate=0.01):
6          self.threshold = threshold
7          self.learning_rate = learning_rate
8          self.weights = np.zeros(no_of_inputs + 1)
9
10     def predict(self, inputs):
11         inputsum = np.dot(inputs, self.weights[1:]) + self.weights[0]
12         if inputsum > 0:
13             activate = 1
14         else:
15             activate = 0
16         return activate
17
18     def train(self, training_inputs, labels):
19         for _ in range(self.threshold):
20             for inputs, label in zip(training_inputs, labels):
21                 prediction = self.predict(inputs)
22                 self.weights[1:] += self.learning_rate * (label - prediction) * inputs
23                 self.weights[0] += self.learning_rate * (label - prediction)
```

FIGURE 10.2 Perceptron.

a specific threshold, then the node in question will fire and send a signal to the next node, the target node (thus the T representing it). The McCulloch–Pitts neuron leads us to what is usually called Hebb's rule. Hebb's rule was discussed previously, but as a reminder, consider this quote from another machine learning textbook:

> Hebb's rule says that the changes in the strength of synaptic connections are proportional to the correlation in the firing of the two connecting neurons. So, if two neurons consistently fire simultaneously, then any connection between them will change in strength, becoming stronger. However, if the two neurons never fire simultaneously, the connection between them will die away. The idea is that if two neurons both respond to something, then they should be connected.[6]

PERCEPTRON

The perceptron is the most basic implementation of a neural network. It is the implementation of a McCulloch–Pitts neuron. Figure 10.2 shows code for a very simple perceptron in Python.

Just 23 lines of code, including blank spaces. The only problem is when you execute this code, it does not print anything out. That is because we did not add any code to print out. We will add that later. For now, notice the three functions in the class Perceptron. We have an initialization function (_init_) that is meant to set initial values. Then we have two functions, one to train and the other to predict. This is a rather simple example; let us look at a more complex example.

The following code uses the mnist dataset. That is the Modified National Institute of Standards and Technology handwriting dataset. This database consists of handwritten numerals. It is often used in machine learning classes to illustrate how to train an algorithm to recognize digits (we will be working with neuroscience-specific datasets a bit later in this chapter). This dataset has 60,000 training images and 10,000 test images. The data was normalized to fit into 28 x 28 pixel boundaries.

```python
import tensorflow as tf
import tensorflow _ datasets as tfds
#the mnist dataset is loaded into two sections
# a test and a training section
(data _ train, data _ test), ds _ info = tfds.load(
'mnist',
split=['train', 'test'],
shuffle _ files=True,
as _ supervised=True,
with _ info=True,
)
def normalize _ image(image, label):
"""Normalizes images: 'uint8' -> 'float32'."""
return tf.cast(image, tf.float32) / 255., label
data _ train = data _ train.map(
normalize _ image, num _ parallel _ calls=tf.data.AUTOTUNE)
data _ train = data _ train.cache()
data _ train  =  data _ train.shuffle(ds _ info.splits['train'].num _
examples)
data _ train = data _ train.batch(128)
data _ train = data _ train.prefetch(tf.data.AUTOTUNE)
data _ test = data _ test.map(
normalize _ image, num _ parallel _ calls=tf.data.AUTOTUNE)
data _ test = data _ test.batch(128)
data _ test = data _ test.cache()
data _ test = data _ test.prefetch(tf.data.AUTOTUNE)
model = tf.keras.models.Sequential([
tf.keras.layers.Flatten(input _ shape=(28, 28)),
tf.keras.layers.Dense(128, activation='relu'),
tf.keras.layers.Dense(10)
])
model.compile(
optimizer=tf.keras.optimizers.Adam(0.001),
```

```
loss=tf.keras.losses.SparseCategoricalCrossentropy(from_
logits=True),
metrics=[tf.keras.metrics.SparseCategoricalAccuracy()],
)
model.fit(
data _ train,
epochs=6,
validation _ data=data _ test,
)
```

When you execute this script, a few things will happen. First, the system may give you errors related to whether or not you have a GPU. If you don't, the algorithm will still run, just not as efficiently. Then the dataset will be loaded. These initial steps are shown in figure 10.3:

Next, the process will continue through six epochs. Notice in the code we set it for six epochs. You can experiment with changing that. The output is shown in figure 10.4.

This is a more robust example of a neural network. You should pause to get comfortable with the code, and executing the script, before you continue further in this chapter.

BACKPROPAGATION

This is an algorithm used in training artificial neural networks, particularly with feed-forward artificial neural networks. Backpropagation is used to compute the gradient of the loss function with respect to the weights. The loss function is also sometimes called a cost or error function. It is a function that is used to represent some cost associated with some activity or event.

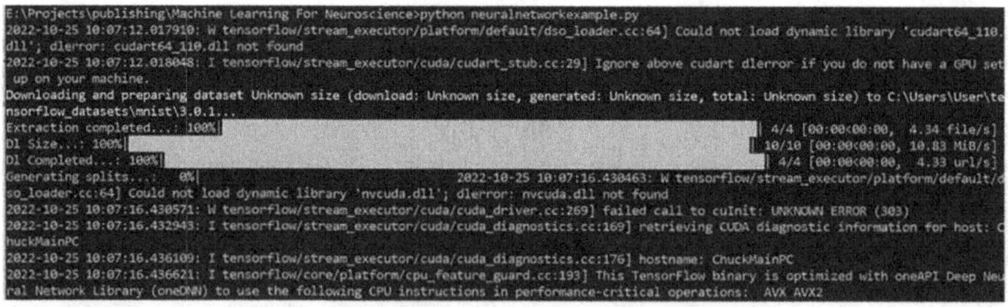

FIGURE 10.3　First neural network.

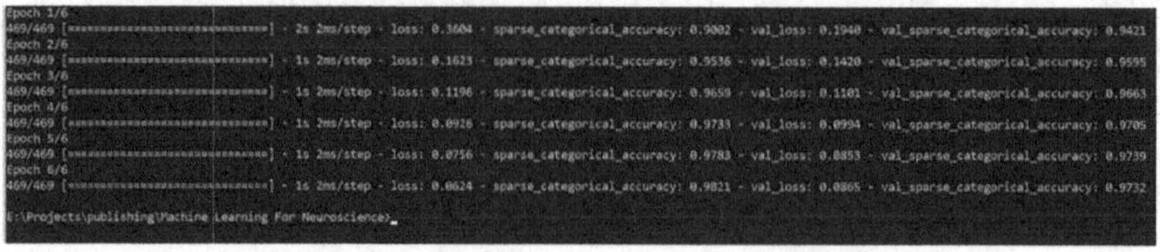

FIGURE 10.4　First neural network output.

The backpropagation algorithm is used to train a neural network using a technique called chain rule. After each forward pass through a network, backpropagation performs a backward pass while adjusting the model's parameters. Normally the parameters adjusted are the weights and biases.

This algorithm was first introduced in the 1970s, but achieved wide recognition due to a 1986 paper in the journal *Nature* written by David Rumelhart, Geoffrey Hitton, and Ronald Williams. It is useful to consider how these authors described backpropagation:

> The procedure repeatedly adjusts the weights of the connections in the network so as to minimize a measure of the difference between the actual output vector of the net and the desired output vector. As a result of the weight adjustments, internal "hidden" units which are not part of the input or output come to represent important features of the task domain, and the regularities in the task are captured by the interactions of these units. The ability to create useful new features distinguishes back-propagation from earlier, simpler methods such as the perceptron-convergence procedure.[7]

Another description from the International Dictionary of Artificial Intelligence[8] could be useful in aiding your understanding of backpropagation:

> A classical method for error propagation when training Artificial Neural Networks (ANNs). For standard backpropagation, the parameters of each node are changed according to the local error gradient. The method can be very slow to converge although it can be improved through the use of methods that slow the error propagation and by batch processing. Many alternate methods such as the conjugate gradient and Levenberg-Marquardt algorithms are more effective and reliable.

Fortunately, the backpropagation algorithm is taken care of by libraries such as TensorFlow. You don't have to implement the details yourself. I say this is fortunate because this algorithm uses partial derivatives, which may be beyond some readers' mathematical skillset.

NORMALIZATION

Normalization is a data-preparation technique that entails altering the values of numeric values in a dataset to a common scale. This is often needed when the values of the dataset have different ranges. There are three different types of normalization used in machine learning:

Input normalization: Normalizing input is quite common. A common example is scaling the pixel values of images (0–255) to values between zero and one.
Batch normalization: This is a normalization that occurs between each layer of the network so that values' mean is zero and their standard deviation is one.

Internal normalization: This normalization is about ensuring that each layer keeps the previous layer's mean and variance.

Many neural network code samples you will see implement normalization at some point. Almost all machine learning scripts dealing with image data will at least implement input normalization.

SPECIFIC VARIATIONS OF NEURAL NETWORKS

Artificial neural networks are widely used (perhaps the most widely used machine learning algorithms). Due to that fact, it should come as no surprise that there are numerous variations of these algorithms. Each has some advantage for some specific application. This section will briefly describe major variations.

RECURRENT NEURAL NETWORKS

The recurrent neural network (RNN) is a common variation of neural networks. These are derived from feedforward neural networks. Recurrent nets are a variation of artificial neural network intended to distinguish patterns in sequences of data. The patterns can be almost anything, but the pattern recognition makes RNNs ideal for identifying handwriting, numbers, time series data from sensors, stock market data, and more. RNNs consider time and sequence. This gives these algorithms a temporal dimension.

The manner in which RNNs function is to have cycles that permit output from some nodes to affect subsequent input into the same nodes. There are several types of RNNs, including:

- Encoder–decoder or sequence-to-sequence RNNs,
- Bidirectional RNNs,
- Recursive RNNs,
- Gated recurrent unit (GRU),
- LSTM RNNs.

The following description of RNNs may help clarify this class of algorithms for you:

A family of neural networks for processing sequential data. RNNs share the same weights across different discrete time steps: each member of the output is a function of the previous members of the output; each member of the output is produced using the same update rule applied to the previous outputs. This recurrent formulation results in the sharing of parameters through a very deep computational graph. RNNs are very powerful dynamic systems for tasks that involve sequential inputs, such as speech and language.[9]

CONVOLUTIONAL NEURAL NETWORKS

A convolutional neural network (CNN) is a neural network with some convolutional layers (and some other layers). A convolutional layer has a number of filters that do convolutional operations. The concept is shown in Figure 10.5.

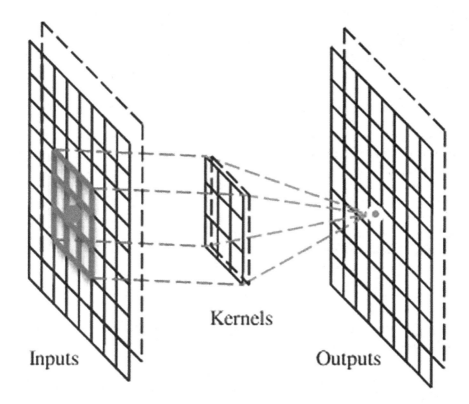

FIGURE 10.5   Convolutional neural network.

The different layers perform different aspects of the learning. Convolutional neural networks (sometimes called ConvNets) are often used with visual imagery. This makes them an ideal candidate for examining diagnostic imagery. Put more formally, convolutional neural networks use a mathematical operation called convolution rather than basic matrix multiplication in at least one of the layers. The architecture of a convolutional neural network will include an input layer, one or more hidden layers, and an output layer. The input is provided as a tensor with a shape. The convolutional neural layer transforms that image to a feature map with a shape. Some sources refer to the feature map as an activation map. The convolutional layers each respond to input only for their own receptive field. This is quite similar to how the brain processes visual imagery.

The following code should help elucidate the convolutional neural network. You may also wish to compare this to the previous code examples, noting similarities and differences.

```
#! /usr/bin/env python
# This is a basic convolutional neural network example.
from _ _ future _ _ import absolute _ import, division, print _
function, unicode _ literals
```

```
import tensorflow as tf
#tensorflow includes a 2D convolutional network, you just
#have to include it.
from tensorflow.keras.layers import Dense, Flatten, Conv2D
from tensorflow.keras import Model
# Load and prepare the MNIST dataset. This dataset is incorporated
with TensorFlow
# the data is split into a training set and a test set
mnist = tf.keras.datasets.mnist
(x_train, y_train), (x_test, y_test) = mnist.load_data()
x_train, x_test = x_train / 255.0, x_test / 255.0
# # Add a channels dimension
x_train = x_train[. . . , tf.newaxis]
x_test = x_test[. . . , tf.newaxis]
# Use tf.data to batch and shuffle the dataset
train_ds = tf.data.Dataset.from_tensor_slices(
        (x_train, y_train)).shuffle(10000).batch(32)
test_ds = tf.data.Dataset.from_tensor_slices(
        (x_test, y_test)).batch(32)
# Build the tf.keras model using the Keras model the layers
will use
# different activation functions
class MyModel(Model):
def __init__(self):
        super(MyModel, self).__init__()
        self.conv1 = Conv2D(32, 3, activation='relu')
        self.flatten = Flatten()
        self.d1 = Dense(128, activation='relu')
        self.d2 = Dense(10, activation='softmax')
def call(self, x):
        x = self.conv1(x)
        x = self.flatten(x)
        x = self.d1(x)
return self.d2(x)
# Create an instance of the model
model = MyModel()
# Choose an optimizer and loss function for training
# there are several other optimizers you can experiment with
# SGD,RMSprop, Adam, Adamax, etc. To see a complet list go to
# www.tensorflow.org/api_docs/python/tf/keras/optimizers
loss_object = tf.keras.losses.SparseCategoricalCrossentropy()
optimizer = tf.keras.optimizers.Adam()
# Select metrics to measure the loss and the accuracy of the
model
# a complete list of metrix can be found here www.tensorflow.org/
api_docs/python/tf/keras/metrics
```

```
train_loss = tf.keras.metrics.Mean(name='train_loss')
train_accuracy = tf.keras.metrics.SparseCategoricalAccuracy(na
me='train_accuracy')
test_loss = tf.keras.metrics.Mean(name='test_loss')
test_accuracy = tf.keras.metrics.SparseCategoricalAccuracy(nam
e='test_accuracy')
# Use tf.GradientTape to train the model.
@tf.function
def train_step(images, labels):
    with tf.GradientTape() as tape:
        predictions = model(images)
        loss = loss_object(labels, predictions)
    gradients = tape.gradient(loss, model.trainable_variables)
    optimizer.apply_gradients(zip(gradients, model.
trainable_variables))
    train_loss(loss)
    train_accuracy(labels, predictions)
@tf.function
def test_step(images, labels):
    predictions = model(images)
    t_loss = loss_object(labels, predictions)
    test_loss(t_loss)
    test_accuracy(labels, predictions)
# you may wish to change the number of epochs to see
# what effect that has
EPOCHS = 5
for epoch in range(EPOCHS):
    for images, labels in train_ds:
        train_step(images, labels)
    for test_images, test_labels in test_ds:
    test_step(test_images, test_labels)
    template = 'Epoch {}, Loss: {}, Accuracy: {}, Test Loss: {},
Test Accuracy: {}'
    print(template.format(epoch+1,
        train_loss.result(),
        train_accuracy.result()*100,
        test_loss.result(),
        test_accuracy.result()*100))
        # Reset metrics for the next epoch
    train_loss.reset_states()
    train_accuracy.reset_states()
    test_loss.reset_states()
    test_accuracy.reset_states()
```

Much of the preceding code should be familiar to you after the previous code samples. A few elements may require some further explanation. One such item is

Epoch 1, Loss: 0.13583704829216003, Accuracy: 95.9800033569336, Test Loss: 0.06109023839235306, Test Accuracy: 97.98999786376953
Epoch 2, Loss: 0.042260274291103851, Accuracy: 98.63166809082031, Test Loss: 0.054545141756534576, Test Accuracy: 98.22000122070312
Epoch 3, Loss: 0.021505581215023994, Accuracy: 99.30833435058594, Test Loss: 0.05119103193283081, Test Accuracy: 98.44999694824219
Epoch 4, Loss: 0.013581412844359875, Accuracy: 99.55166625976562, Test Loss: 0.04984024912118912, Test Accuracy: 98.50999450683594
Epoch 5, Loss: 0.008280626498162746, Accuracy: 99.7266616821289, Test Loss: 0.0580204613506794, Test Accuracy: 98.3699951171875

FIGURE 10.6   CNN output.

SparseCategoricalCrossentropy. This is a property of Keras that is used when there are two or more label classes, in order to compute the cross-entropy loss between the labels and predictions.

When you execute this script, you may first get familiar GPU error messages if your computer does not have a suitable GPU. Then you will see the epochs, each with an accuracy shown. This is shown in Figure 10.6.

Note that the test accuracy improved to a point, then began to level off and even decrease slightly. This is an important fact to keep in mind. The issue is not simply to do as many epochs as you possibly can. There is a point where this will simply generate diminishing returns. However, experimenting with different activation and optimization functions can be quite useful.

Given how common convolutional neural networks are, we will review a second code sample. This one is used to detect pituitary brain tumors based on imaging. The code is more detailed and should aid you in understanding CNNs. The data can be found on Github at https://github.com/sartajbhuvaji/brain-tumor-classification-dataset

```
import numpy as np
import pandas as pd # used for reading CSV data files
import os #needed for navigating file system
#The path is based on having downloaded the data to a
#subfolder of the directory this script is in and the
#folder was named 'braintumordataset'
for dirname, _ , filenames in os.walk('/BrainTumorDataSet'):
    for filename in filenames:
        print(os.path.join(dirname, filename))
#keras is used for our CNN
import keras
from keras.models import Sequential
from keras.layers import Conv2D,Flatten,Dense,MaxPooling2D,Dro
pout
from sklearn.metrics import accuracy_score
import io
from PIL import Image
import tqdm
from sklearn.model_selection import train_test_split
import cv2
from sklearn.utils import shuffle
import tensorflow as tf
import matplotlib.pyplot as plt
```

```
xtrainarray = []
ytrainarray = []
imgsize = 150
labels        =        ['glioma _ tumor','meningioma _ tumor','no _
tumor','pituitary _ tumor']
for i in labels:
    folderPath = os.path.join('BrainTumorDataSet/Training/',i)
    for j in os.listdir(folderPath):
        img = cv2.imread(os.path.join(folderPath,j))
        img = cv2.resize(img,(imgsize,imgsize))
        xtrainarray.append(img)
        ytrainarray.append(i)

for i in labels:
    folderPath = os.path.join('BrainTumorDataSet/Testing/',i)
    for j in os.listdir(folderPath):
        img = cv2.imread(os.path.join(folderPath,j))
        img = cv2.resize(img,(imgsize,imgsize))
        xtrainarray.append(img)
        ytrainarray.append(i)
xtrainarray = np.array(xtrainarray)
ytrainarray = np.array(ytrainarray)
xtrainarray,ytrainarray    =    shuffle(xtrainarray,ytrainarray,ran
dom _ state=101)
xtrainarray.shape
xtrainarray,X _ test,ytrainarray,y _ test = train _ test _ split(xt
rainarray,ytrainarray,test _ size=0.1,random _ state=101)
ytrainarray _ new = []
for i in ytrainarray:
    ytrainarray _ new.append(labels.index(i))
ytrainarray=ytrainarray _ new
ytrainarray = tf.keras.utils.to _ categorical(ytrainarray)
y _ test _ new = []
for i in y _ test:
    y _ test _ new.append(labels.index(i))
y _ test=y _ test _ new
y _ test = tf.keras.utils.to _ categorical(y _ test)
# you should experiment with different activation functions
# Tanh, relu, relu6, gelu, etc.
model = Sequential()
model.add(Conv2D(32,(3,3),activation              =              'relu6',inp
ut _ shape=(150,150,3)))
model.add(Conv2D(64,(3,3),activation= 'relu6'))
model.add(MaxPooling2D(2,2))
model.add(Dropout(0.3))
model.add(Conv2D(64,(3,3),activation='relu6'))
```

```
model.add(Conv2D(64,(3,3),activation='relu6'))
model.add(Dropout(0.3))
model.add(MaxPooling2D(2,2))
model.add(Dropout(0.3))
model.add(Conv2D(128,(3,3),activation='relu6'))
model.add(Conv2D(128,(3,3),activation='relu6'))
model.add(Conv2D(128,(3,3),activation='relu6'))
model.add(MaxPooling2D(2,2))
model.add(Dropout(0.3))
model.add(Conv2D(128,(3,3),activation='relu'))
model.add(Conv2D(256,(3,3),activation='relu'))
model.add(MaxPooling2D(2,2))
model.add(Dropout(0.3))
model.add(Flatten())
model.add(Dense(512,activation = 'relu'))
model.add(Dense(512,activation = 'relu'))
model.add(Dropout(0.3))
model.add(Dense(4,activation='softmax'))
model.summary()
model.compile(loss='categorical _ crossentropy',optimizer='Adam',
metrics=['accuracy'])
#5 epochs is just so this will run quickly. In real applications
you will often
# use more.
history     =     model.fit(xtrainarray,ytrainarray,epochs=5,validat
ion _ split=0.1)
acc = history.history['accuracy']
val _ acc = history.history['val _ accuracy']
epochs = range(len(acc))
fig = plt.figure(figsize=(14,7))
plt.plot(epochs,acc,'r',label="Training Accuracy")
plt.plot(epochs,val _ acc,'b',label="Validation Accuracy")
plt.legend(loc='upper left')
plt.show()
loss = history.history['loss']
val _ loss = history.history['val _ loss']
epochs = range(len(loss))
fig = plt.figure(figsize=(14,7))
plt.plot(epochs,loss,'r',label="Training loss")
plt.plot(epochs,val _ loss,'b',label="Validation loss")
plt.legend(loc='upper left')
plt.show()
img = cv2.imread('/BrainTumorDataSet/Training/pituitary _ tumor/p
(107).jpg')
img = cv2.resize(img,(150,150))
img _ array = np.array(img)
```

```
img _ array.shape
img _ array = img _ array.reshape(1,150,150,3)
img _ array.shape
from tensorflow.keras.preprocessing import image
img = image.load _ img('/BrainTumorDataSet/Training/pituitary _
tumor/p (107).jpg')
plt.imshow(img,interpolation='nearest')
plt.show()
a=model.predict(img _ array)
indices = a.argmax()
indices
```

Notice the line of code model.summary(), model.summary will display the choices you made in your model on the screen. You could omit this line and the script would still work. Or you may try including this line in other scripts where you build a keras. model. When you execute this script, several images will be displayed: these are shown in figures 10.7 through 10.10.

Layer (type)	Output Shape	Param #
conv2d (Conv2D)	(None, 148, 148, 32)	896
conv2d_1 (Conv2D)	(None, 146, 146, 64)	18496
max_pooling2d (MaxPooling2D)	(None, 73, 73, 64)	0
dropout (Dropout)	(None, 73, 73, 64)	0
conv2d_2 (Conv2D)	(None, 71, 71, 64)	36928
conv2d_3 (Conv2D)	(None, 69, 69, 64)	36928
dropout_1 (Dropout)	(None, 69, 69, 64)	0
max_pooling2d_1 (MaxPooling 2D)	(None, 34, 34, 64)	0
dropout_2 (Dropout)	(None, 34, 34, 64)	0
conv2d_4 (Conv2D)	(None, 32, 32, 128)	73856
conv2d_5 (Conv2D)	(None, 30, 30, 128)	147584
conv2d_6 (Conv2D)	(None, 28, 28, 128)	147584
max_pooling2d_2 (MaxPooling 2D)	(None, 14, 14, 128)	0

FIGURE 10.7   CNN-model summary.

```
Epoch 1/5
83/83 [==============================] - 98s 1s/step - loss: 1.4175 - accuracy: 0.2838 - val_loss: 1.3652 - val_accuracy: 0.2891
Epoch 2/5
83/83 [==============================] - 98s 1s/step - loss: 1.3617 - accuracy: 0.2762 - val_loss: 1.3630 - val_accuracy: 0.2721
Epoch 3/5
83/83 [==============================] - 103s 1s/step - loss: 1.3593 - accuracy: 0.2872 - val_loss: 1.3633 - val_accuracy: 0.2721
Epoch 4/5
83/83 [==============================] - 99s 1s/step - loss: 1.3612 - accuracy: 0.2720 - val_loss: 1.3622 - val_accuracy: 0.2891
Epoch 5/5
83/83 [==============================] - 101s 1s/step - loss: 1.3597 - accuracy: 0.2788 - val_loss: 1.3648 - val_accuracy: 0.2721
```

FIGURE 10.8   CNN-epoch results.

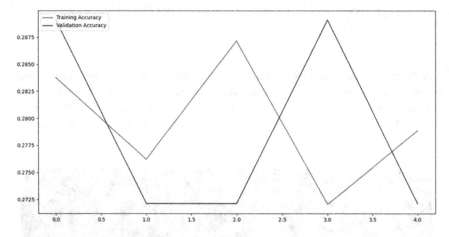

FIGURE 10.9   CNN accuracy.

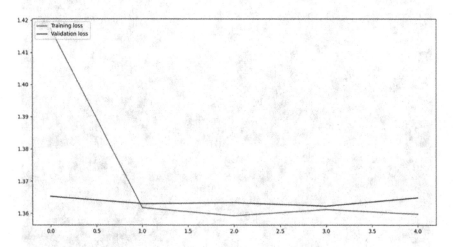

FIGURE 10.10   CNN loss.

Autoencoder

This is a variation of artificial neural network that is used to learn how to properly code unlabeled data. The algorithm validates its results by trying to regenerate the input data from the encoders. The encoding is simply a representation of the data.

Autoencoders are often used for dimensionality reduction. An autoencoder has two main parts: an encoder that maps the message to a code, and a decoder that reconstructs the message from the code. An optimal autoencoder would perform as close to perfect reconstruction as possible. How close to perfection the autoencoder performs in actuality is determined by a quality function.

As you can probably guess, there are numerous variations of autoencoders. There are sparse autoencoders (SAE), denoising autoencoders (DAE), and others. Sparse autoencoders intentionally introduce bottlenecks, but rather than do so via reducing the number of nodes, the SAE instead penalizes particular activations of nodes with a layer. Denoising autoencoders work by introducing a minor deviation into the input data, but maintaining the unmodified data as the target output. The model needs to learn to denoise (remove deviations) from the data.

The following code is a basic autoencoder. This code again uses the MINST database because of its easy availability.

```python
from keras.layers import Dense,Conv2D,MaxPooling2D,UpSampling2D
from keras import Input, Model
from keras.datasets import mnist
import numpy as nump
import matplotlib.pyplot as plt
#first build the model
encoding _ dimension = 15
inumput _ img = Input(shape=(784,))
# encoded representation of input
encoded    =    Dense(encoding _ dimension,    activation='relu')
(input _ img)
# decoded representation of code
decoded = Dense(784, activation='sigmoid')(encoded)
# Model which take input image and shows decoded images
autoencoder = Model(input _ img, decoded)
# Now build the encoder and decoder
encoder = Model(input _ img, encoded)
# Creating a decoder model
encoded _ input = Input(shape=(encoding _ dimension,))
# last layer of the autoencoder model
decoder _ layer = autoencoder.layers[-1]
# decoder model
decoder = Model(encoded _ input, decoder _ layer(encoded _ input))
#compile the model. Note the optimizer we are using
#you can try this with a different optimizer.
autoencoder.compile(optimizer='adam',
loss='binary _ crossentropy')
#now we load the minst data
(x _ train, y _ train), (x _ test, y _ test) = mnist.load _ data()
x _ train = x _ train.astype('float32') / 255.
x _ test = x _ test.astype('float32') / 255.
```

```
x _ train = x _ train.reshape((len(x _ train), nump.prod(x _ train.
shape[1:])))
x _ test   =   x _ test.reshape((len(x _ test),   nump.prod(x _ test.
shape[1:])))
print(x _ train.shape)
print(x _ test.shape)
```

When executed, it will simply output the following:

(60000, 784)
(10000, 784)

The following quote might help elucidate the autoencoder:

> The combination of an encoder function that converts the input data into a different representation, and a decoder function that converts the new representation back into the original format. Autoencoders are trained to preserve as much information as possible when an input is run through the encoder and then the decoder, but are also trained to make the new representation have various nice properties. Different kinds of autoencoders aim to achieve different kinds of properties.[10]

SPIKING NEURAL NETWORK

Another variation of the artificial neural network is one that incorporates the concept of time. With a spiking neural network, information is not transmitted at every propagation cycle. Rather, information is only transmitted when a particular threshold is met. This mimics the manner in which biological neurons function. Spiking neural networks are based on the Hodgkin–Huxley model of biological networks. This model describes how action potentials are initiated and propagated to the next neuron. The development of this model earned Alan Hodgkin and Andrew Huxley the 1963 Nobel Prize in Physiology or Medicine.

The concept is for the nodes in a given layer to not test for activation in every iteration of propagation. The nodes only test for activation when their potentials reach a specific value. This is different than typical multilayer perceptrons wherein neurons test for propagation every iteration.

DEEP NEURAL NETWORKS

While the terminology may indicate that this is a substantial variation on the neural network concept, it is not. A deep neural network (DNN) simply has more hidden layers. This is shown in Figure 10.11.

In Figure 10.11, the connections in the hidden layers are not shown. This is because the hidden layers may have all nodes connected to all other nodes, or only some nodes connected to other nodes. A deep neural network can also be a feedforward network or a recurrent network. Each layer may even use a different activation

n Hidden Layers

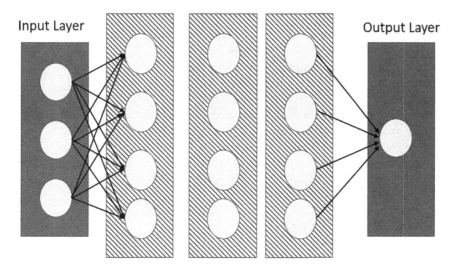

FIGURE 10.11   Deep neural network.

function. Deciding how many layers and what activation functions should be used is not a straightforward process. It requires some level of experience, and a bit of experimentation to find the right combination for a given problem.

NEUROSCIENCE EXAMPLE CODE

The following is an example of Python code to recognize brain tumors from diagnostic imaging data. You will need to do a few things to make this code function. First, if you have not previously done so, you will need to install two items:

```
pip install opencv-python
pip install scikit-learn
```

The first allows you to open comma-delimited files. The second allows you to use the scikit-learn package. Then you need to download the dataset. There are a number of datasets freely available on websites like Kaggle.com. This dataset comes from www.kaggle.com/datasets/preetviradiya/brian-tumor-dataset

You will download it to the same folder you have your Python script in. The code is shown here:

```
import cv2
import os
import tensorflow as tf
from tensorflow.keras import models
```

```
from tensorflow.keras import layers
from tqdm import tqdm
from sklearn.model_selection import train_test_split
import numpy as np
# there are two data groups in the set, health and tumor
loc0 = 'Brain Tumor Data Set/Healthy'
loc1 = 'Brain Tumor Data Set/Brain Tumor'
# set up an array for your features
features = []
#tqdm is used to create a progress bar, definitely makes your
script more user friendly
for img in tqdm(os.listdir(loc0)):
    f = cv2.imread(os.path.join(loc0,img)) #read in healthy data
    f = cv2.cvtColor(f,cv2.COLOR_BGR2GRAY)
    f = cv2.resize(f,(70,70))
    features.append(f)
for img in tqdm(os.listdir(loc1)):
    f = cv2.imread(os.path.join(loc1,img)) # read in tumor data
    f = cv2.cvtColor(f,cv2.COLOR_BGR2GRAY)
    f = cv2.resize(f,(70,70))
    features.append(f)

#now we have an array for the labels.
labels = []
for img in tqdm(os.listdir(loc0)):
    labels.append(0)
for img in tqdm(os.listdir(loc1)):
    labels.append(1)
X = np.array(features)
X.shape
Y = np.array(labels)
Y.shape
Xt = X.reshape(4600, 4900)
Yt = Y.reshape(4600,1)
xtrain,xtest,ytrain,ytest   =   train_test_split(Xt,Yt,train_
size=0.75)
ytrainC = tf.keras.utils.to_categorical(ytrain)
ytestC = tf.keras.utils.to_categorical(ytest)
# we are using sequential models with specific activation func-
tions. Once you are comfortable
# with this script you can experiment with different models and
different activation functions
model = models.Sequential()
model.add(layers.Dense(300,      activation='relu',      input_
dim = xtrain.shape[1]))
model.add(layers.Dense(200,activation='relu'))
```

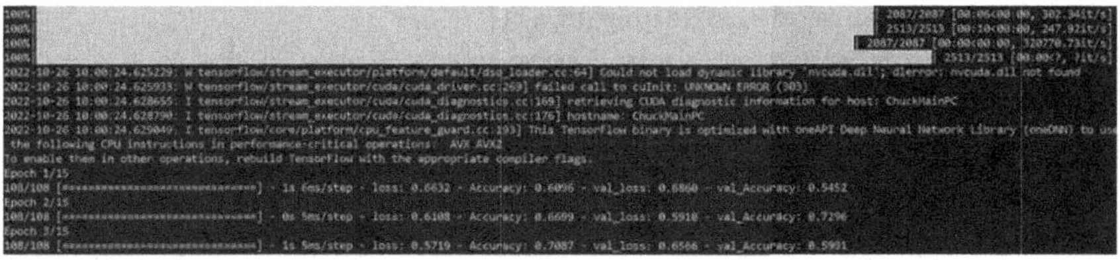

FIGURE 10.12   Using brain-tumor imagery.

```
model.add(layers.Dense(200,activation='relu'))
model.add(layers.Dense(100,activation='relu'))
model.add(layers.Dense(2,activation='sigmoid'))
xtrainN = xtrain/xtrain.max()
xtestN = xtest/xtest.max()
SGD = tf.keras.optimizers.SGD(0.1)
model.compile(loss = 'categorical _ crossentropy',
              optimizer = SGD,
              metrics=['Accuracy'])
model.fit(xtrainN,ytrainC,epochs=15,validation_data=(xtestN,ytestC))
```

When you execute the code you should see something like what is shown in Figure 10.12.

Note the progress bars as data is being read into the script. Then you see the epochs begin, with each having a particular accuracy.

SUMMARY

In this chapter we have covered the basic concepts of neural networks, and the terminology needed to work with artificial neural networks, as well as exemplary scripts. Activation functions, optimization functions, and models are all concepts you should be comfortable with after reading this chapter. You should be very familiar with these scripts before you proceed to the next chapter.

EXERCISES

LAB 1: BASIC TENSORFLOW

```
#!/usr/bin/python
# this is a basic TensorFlow project
import tensorflow as tf
mnist = tf.keras.datasets.mnist
# load the data set
(x _ train, y _ train), (x _ test, y _ test) = mnist.load _ data()
x _ train, x _ test = x _ train / 255.0, x _ test / 255.0
#Build the TensorFlow tf.keras.Sequential model by stacking
layers.
```

```
model = tf.keras.models.Sequential([
tf.keras.layers.Flatten(input _ shape=(28, 28)),
tf.keras.layers.Dense(128, activation='relu'),
tf.keras.layers.Dropout(0.2),
tf.keras.layers.Dense(10)
])
#For each example the model returns a vector of "logits"
#or "log-odds" scores, one for each class.
predictions = model(x _ train[:1]).numpy()
predictions
#The TensorFlow tf.nn.softmax function converts these logits
#to "probabilities" for each class:
tf.nn.softmax(predictions).numpy()
#The losses.SparseCategoricalCrossentropy loss takes a vector
of logits
#and a True index and returns a scalar loss for each example.
loss _ fn = tf.keras.losses.SparseCategoricalCrossentropy(from _
logits=True)
loss _ fn(y _ train[:1], predictions).numpy()
model.compile(optimizer='adam',
loss=loss _ fn,
metrics=['accuracy'])
model.fit(x _ train, y _ train, epochs=5)
#The Model.evaluate method checks the model's performance
model.evaluate(x _ test, y _ test, verbose=2)
Now run it
```

```
D:\MLPython>python firstexample.py
2020-08-05 15:40:04.524759: W tensorflow/stream_executor/platform/default/dso_loader.cc:55] Could not load dynamic libr
ary 'cudart64_101.dll'; dlerror: cudart64_101.dll not found
2020-08-05 15:40:04.561741: I tensorflow/stream_executor/cuda/cudart_stub.cc:29] Ignore above cudart dlerror if you do
not have a GPU set up on your machine.
Downloading data from https://storage.googleapis.com/tensorflow/tf-keras-datasets/mnist.npz
11493376/11490434 [==============================] - 2s 0us/step
2020-08-05 15:40:31.075274: W tensorflow/stream_executor/platform/default/dso_loader.cc:55] Could not load dynamic libr
ary 'nvcuda.dll'; dlerror: nvcuda.dll not found
2020-08-05 15:40:31.079931: E tensorflow/stream_executor/cuda/cuda_driver.cc:313] failed call to cuInit: UNKNOWN ERROR
(303)
2020-08-05 15:40:31.084328: I tensorflow/stream_executor/cuda/cuda_diagnostics.cc:169] retrieving CUDA diagnostic infor
mation for host: WIN-7EP9LVQV3O7
2020-08-05 15:40:31.086703: I tensorflow/stream_executor/cuda/cuda_diagnostics.cc:176] hostname: WIN-7EP9LVQV3O7
2020-08-05 15:40:31.097392: I tensorflow/core/platform/cpu_feature_guard.cc:143] Your CPU supports instructions that th
is TensorFlow binary was not compiled to use: AVX2
2020-08-05 15:40:31.291817: I tensorflow/compiler/xla/service/service.cc:168] XLA service 0x2be43b16a80 initialized for
platform Host (this does not guarantee that XLA will be used). Devices:
2020-08-05 15:40:31.296403: I tensorflow/compiler/xla/service/service.cc:176]   StreamExecutor device (0): Host, Defaul
```

```
Epoch 1/5
1875/1875 [==============================] - 2s 864us/step - loss: 0.2973 - accuracy: 0.9132
Epoch 2/5
1875/1875 [==============================] - 1s 782us/step - loss: 0.1461 - accuracy: 0.9565
Epoch 3/5
1875/1875 [==============================] - 1s 772us/step - loss: 0.1101 - accuracy: 0.9668
Epoch 4/5
1875/1875 [==============================] - 1s 796us/step - loss: 0.0901 - accuracy: 0.9716
```

LAB 2: PERCEPTRON

```
import numpy as np
X = np.array([
    [-2,4,-1],
    [4,1,-1],
    [1, 6, -1],
    [2, 4, -1],
    [6, 2, -1],
])
y = np.array([-1,-1,1,1,1])
def perceptron_sgd(X, Y):
    w = np.zeros(len(X[0]))
    eta = 1
    epochs = 20
    for t in range(epochs):
        for i, x in enumerate(X):
            if (np.dot(X[i], w)*Y[i]) <= 0:
                w = w + eta*X[i]*Y[i]
    return w
w = perceptron_sgd(X,y)
print(w)
```

NOTES

1. https://developers.google.com/machine-learning/glossary
2. www.tensorflow.org/api_docs/python/tf/keras/optimizers
3. https://blogs.nvidia.com/blog/2021/08/16/what-is-a-machine-learning-model/
4. https://tfhub.dev/
5. https://link.springer.com/content/pdf/10.1007/BF02478259.pdf
6. *Machine learning: an algorithmic perspective*, Second Edition. Chapman & Hall/Crc Machine Learning & Pattern Recognition.
7. www.nature.com/articles/323533a0
8. https://citeseerx.ist.psu.edu/viewdoc/download?doi=10.1.1.375.8194&rep=rep1&type=pdf
9. Goodfellow, I., Bengio, Y. and Courville, A. *Deep learning* (p. 373). MIT Press. LeCun, Y., Bengio, Y. and Hinton, G. 2015. Deep learning. *Nature*, 521, 436–444.
10. Goodfellow, I., Bengio, Y. and Courville, A. *Deep learning* (p. 8). MIT Press.

11 More with ANN

INTRODUCTION

Artificial neural networks are among the most common machine learning algorithms. This means there are also many variations. Some of these algorithms are minor variations on the basic concept of neural networks, while others are substantial variations. It is important to understand these algorithms and how they can be used in neuroscience research.

MORE ACTIVATION FUNCTIONS

In chapter 10, several common activation functions were described. There are other activation functions that, while not as common as those in chapter 10, are still worth briefly describing. As you work through various code samples in this chapter, it is important to determine which activation function is most effective for a given application. Put another way, there are many different activation functions because there is no single "best" activation function. Given variations in datasets, objectives, and other parameters, you may find substantial differences by simply changing the activation function utilized.

SELU

Scaled Exponential Linear Unit (SELU) was first described in a paper entitled "Self-Normalizing Neural Networks."[1] SELU has self-normalization; this is what differentiates it from RELU and similar activation functions. Normalization was discussed in chapter 10, but is briefly summarized here: normalization is a data-preparation technique that entails altering the values of numeric values in a dataset to a common scale.

Input normalization: Normalizing input is quite common. A common example is scaling the pixel values of images (0–255) to values between zero and one.

Batch normalization: This is a normalization that occurs between each layer of the network, so that values' mean is zero and their standard deviation is one.

Internal normalization: This normalization is about ensuring that each layer keeps the previous layer's mean and variance. SELU focuses on internal normalization.

The problem being solved by self-normalization, and thus by SELU, is vanishing gradients. When using backpropagation (described in chapter 10), the gradients can get too small, then the weights no longer adjust and learning stops. The SELU function is described in equation 11.1.

$$f(\alpha, x) = \lambda \begin{cases} \alpha\left(e^x - 1\right) & \text{for } x < 0 \\ x & \text{for } x \geq 0 \end{cases} \qquad \text{(eq 11.1)}$$

In equation 11.1, the λ α (lambda and alpha) symbols represent scaling factors. TensorFlow and PyTorch uses $\lambda = 1.0507$. The lambda is sometimes presented in a longer form of $1.0507009873554804934193349852946$. The alpha value is approximately 1.67326. The formula given basically states that if $x \geq 0$, then multiply x by lambda. If $x < 0$, then we take the exponential value of ex–1 and multiply that by lambda and alpha. A simple Python SELU function is shown here:

```
Import numpy as np
def SELU(x, lambda = 1.0507, alpha = 1.6732):
    if x >= 0:
        return lambda * x
    else:
        return lambda * alpha * (np.exp(x) - 1)
```

However, you won't need to write your own SELU function, as SELU is part of the Keras api. You just use:

tf.keras.activations.selu(x)

If you have been using RELU (from chapter 10) and encounter the "dying RELU" problem, then SELU is an alternative you should consider. It is also common to find that input normalization is necessary for effective image recognition. Therefore, if your problem involves image recognition, such as identifying tumors in an MRI image, SELU can be more effective than RELU. Also, if you are working with deep neural networks, then SELU is preferred over RELU.

Silu

Sigmoid-weighted linear units are another interesting activation function. As the name suggests, this begins with the sigmoid function. This function is shown in equation 11.2.

$$\sigma(x) = \frac{1}{1 + e^{-x}} \qquad \text{(eq. 11.2)}$$

When using SILU, the activation function for the input k is computed by the sigmoid function multiplied by its input, shown in equation 11.3.

$$a(k) = k\sigma(k) \qquad \text{(eq. 11.3)}$$

The sigmoid linear unit (SiLU), sometimes called the sigmoid-weighted linear unit, is an activation function that uses the sigmoid function and multiplication. As you can see in equation 11.2, the activation of the SiLU is computed by the sigmoid function multiplied by its input. Sigmoid is generally recommended for reinforcement-learning scenarios. SiLU is often used in hidden layers in deep learning networks, but is commonly not recommended for the input and output layers. There are research papers that delve deeply into the SiLU activation function.[2]

There are variations of SiLU. One common variation is derivative of the sigmoid-weighted linear unit (dSiLU). The dSiLU is often described as a steeper and "overshooting" version of the SiLU function. More formally speaking, it is the gradient of the SiLU function. It is often used for gradient-descent updates of weighting in neural networks.

SWISH

This activation function is far less common than the others we have discussed. The formula for swish has been described with minor differences in various sources. It is a modified version of SiLU. Some libraries, such as PyTorch, use the terms SiLU and Swish interchangeably,[3] but this is not accurate. The Swish algorithm makes a minor change to the SiLU function by adding in a trainable parameter β. This changes the function to what is shown in equation 11.4.

$$swish(x) = x\, sigmoid(\beta x) = \frac{x}{1 + e^{-\beta x}} \qquad \text{(eq. 11.4)}$$

Comparing Swish to the more common RELU activation function, RELU is a piecewise linear function whereas Swish is a smooth continuous function. RELU moves all negative weights to zero, while Swish allows negative weights to be propagated. This function is part of the Keras api, you just use

tf.keras.activations.swish(x)

Given that so many of these activation functions are incorporated into Keras, it is quite easy not only to implement a given activation function, but to quickly change activation functions. This allows you a wide range of activation functions you can quickly experiment with to determine which is the most effective for a given application.

SOFTSIGN

Softsign is an alternative to tanh (tanh was discussed in chapter 10). The formula for tanh is shown in equation 11.5.

$$\tanh(x) = \frac{e^x - e^{-x}}{e^x + e^{-x}} \qquad \text{(eq. 11.5)}$$

The Softsign activation function is shown in equation 11.6.

$$f(z) = \left(\frac{x}{|x|+1} \right) \tag{eq. 11.6}$$

This function is part of the Keras api, you just use

tf.keras.activations.softsign(x)

While tanh and Softsign functions are closely related the manner in which their functions converge is different. The tanh activation function converges exponentially, whereas Softsign converges polynomially. Both tanh and Softsign product outputs are in the range of −1 to +1.

ALGORITHMS

In this chapter, we will examine variations of artificial neural networks. As was discussed in chapter 10, there are numerous such variations. Each having its own strengths and weaknesses. It is important to understand these algorithms in order to select an appropriate algorithm for a particular purpose. The information from that chapter is repeated here.

SPIKING NEURAL NETWORKS

There are actually a number of variations of the SNN, but all have some common properties. The most important is that all spiking neural networks integrate time into the operation of the algorithm. The basic concept of an SNN is simple. Each node/neuron does not transmit data at each cycle/epoch. Rather, when the node reaches a particular threshold of input, it fires. This more accurately mimics the behavior of biological neurons.

Perhaps the most common SNN model is the leaky integrate-and-fire model. This model, as a mathematical function, was introduced in chapter 8 in reference to computational neuroscience. This leaky integrate-and-fire model was developed in the 20th century by neuroscientist Louis Lapicquie. This model uses the term leak to reflect the diffusions of ions through the membrane. This is related to the alternative model, aptly named the non-leaky integratre-and-fire model. The basic model for diffusion of ions through the membrane is shown in equation 11.7:

$$C_m \frac{dV_m(t)}{dt} = I(t) - \frac{V_m(t)}{R_m} \tag{eq. 11.7}$$

Again, this formula includes a derivative with respect to time. In equation 11.7, V_m is the voltage across the cell membrane, t represents time, and R_m is the resistance of the cell membrane. I, as is typical in electronic diagrams, represents current. The concept of an artificial SNN is that at each point in time, each node/neuron has some value that is analogous to the membrane potential of biological neurons. That value can change over time. If and when that value exceeds some threshold, then the node/neuron will send an impulse to the next downstream node/neuron, and the value equivalent to a membrane potential will drop.

The artificial SNN, which is our focus, can be used in most of the same applications as any artificial neural network. However, it has often been noted that spiking neural networks can have a slightly lower accuracy rate than other artificial neural networks. This indicates that one should only use the SNN when its functionality provides some specific advantage. There is a simulator online that will allow you to work with spiking neural networks.[4]

SNNs are often divided into subgroups. These include feedforward neural networks, recurrent neural networks, and hybrid neural networks that have some nodes feedforward and some recurrent. The concepts of feedforward and recurrent networks were discussed in chapter 10. There is Python code on GitHub that emulates a leaky integrate-and-fire model.[5] Also, the SpykeTorch[6] library includes spiking neural networks.

LIQUID STATE MACHINE

LSM is a type of spiking neural network. As discussed in the preceding section, spiking neural networks (SNN) incorporate time into the model. Neurons don't necessarily fire at each propagation cycle, but fire only when a particular membrane potential reaches a particular value, thus mimicking neurons.

The term liquid, in the current context, refers to data flow. LSMs are sometimes used to describe brain operations. A liquid state machine typically has a large number of nodes. Each node will receive inputs that vary both in strength and temporally. Inputs can come from external sources or from other nodes. The is due to the fact that the LSM is recurrent, thus having nodes connecting and transmitting to other nodes.

LONG SHORT-TERM MEMORY NEURAL NETWORKS

A variation of recurrent neural networks (RNN):

> Recurrent nets are a type of artificial neural network designed to recognize patterns in sequences of data, such as text, genomes, handwriting, the spoken word, or numerical times series data emanating from sensors, stock markets and government agencies. These algorithms take time and sequence into account, they have a temporal dimension.[7]

The key to LSTMs is the cell state. LSTMs also have this chain-like structure, but the repeating module has a different structure. Instead of having a single neural network layer, there are four, interacting in a very particular way:

> The cell state runs straight down the entire chain, with only some minor linear interactions. The LSTM has the ability to remove or add information to the cell state, regulated by gates. The sigmoid layer outputs numbers between zero and one, describing how much of each component should be let through.

A value of zero means nothing gets through while a value of one allows everything through. An LSTM has three of these gates, to protect and control the cell state.[8]

The standard structure of an LSTM is basically four neural networks and a number of memory blocks called cells, put together in a chain. A single unit of LSTM has a cell (memory block), an input gate, an output gate, and a forget gate.

The input gate decides which of the input values should be stored in memory. The sigmoid function is used to determine whether to allow 0 or 1 values through. And the tanh function assigns weight to the data using a scale of −1 to 1. The output gate uses the sigmoid function to decide whether to allow 0 or 1 values through. The output gate also uses the tanh function to decide which values are allowed to pass through 0, 1. The tanh function also assigns weight to the values provided. The forget gate finds the details that should be removed from the block (i.e., forgotten). The output gate uses a sigmoid function to make this determination. For each number in the cell state, the output gate examines the preceding state and the content input and produces a number between 0 (forget this) and 1 (remember this).

The following quote may assist you in understanding LSTM:

> Long Short-Term Memory (LSTM) networks are a type of recurrent neural network capable of learning order dependence in sequence prediction problems. This is a behavior required in complex problem domains like machine translation, speech recognition, and more. LSTMs are a complex area of deep learning. It can be hard to get your hands around what LSTMs are, and how terms like bidirectional and sequence-to-sequence relate to the field.[9]

BOLTZMANN MACHINE

A Boltzmann machine is a generative unsupervised algorithm. The name is derived from Boltzmann distribution (also called Gibbs distribution), which is a probability distribution that provides the probability that a given system will be in a particular state, as a function of the state's energy. This algorithm uses the process of learning a probability distribution from a given dataset to make inferences about new data sets. As is common for neural networks, Boltzmann machines have an input layer (sometimes called the visible layer) then one or more hidden layers. The nodes in a Boltzmann machine are connected not only to other nodes in other layers but also to neurons within the same layer. This is shown in Figure 11.1 (the black nodes are output nodes, and the empty circles are the hidden variables).

The primary goal of a Boltzmann machine is to optimize the solution of a given problem. The nodes each make binary decisions. The name of this algorithm comes from 19th-century physicist Ludwig Boltzmann who, among other things, established the Boltzmann constant, which describes the proportionality factor relating the average relative kinetic energy of particles in a gas with the thermodynamic temperature of the gas. However, the actual algorithm was developed by Geoffrey Hinton.

Boltzmann machines optimize weights related to a given problem and create a mapping from the attributes in the data. Boltzmann machines can be used to

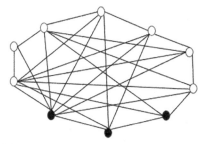

FIGURE 11.1 Boltzmann machine.

determine the structures or patterns that may exist in that data. The following code uses a Boltzmann machine to recognize handwritten digits. While that is not a neuroscience problem, it is a common problem to learn various machine learning algorithms with.

```python
import numpy as np
import torch
import torch.utils.data
import torch.nn as nn
import torch.nn.functional as func
import torch.optim as optimize
from torch.autograd import Variable
from torchvision import datasets, transforms
from torchvision.utils import make_grid, save_image
import matplotlib.pyplot as plt
batch_size = 64
#we will use the MNIST dataset of handwritten numbers
train_loader = torch.utils.data.DataLoader(
datasets.MNIST('./data',
    train=True,
    download = True,
    transform = transforms.Compose(
        [transforms.ToTensor()])
    ),
    batch_size=batch_size
)
#load the data
test_loader = torch.utils.data.DataLoader(
datasets.MNIST('./data',
    train=False,
    transform=transforms.Compose(
    [transforms.ToTensor()])
    ),
```

```
     batch _ size=batch _ size)
#this class is the key so study it well
class RBM(nn.Module):
   def _ _ init _ _ (self,
               n _ vis=784,
               n _ hin=500,
               k=5):
       super(RBM, self)._ _ init _ _ ()
       self.W = nn.Parameter(torch.randn(n _ hin,n _ vis)*1e-2)
       self.v _ bias = nn.Parameter(torch.zeros(n _ vis))
       self.h _ bias = nn.Parameter(torch.zeros(n _ hin))
       self.k = k

   def sample _ from _ p(self,p):
       return func.relu(torch.sign(p - Variable(torch.rand(p.
size())))))

   def v _ to _ h(self,v):
       p _ h = func.sigmoid(F.linear(v,self.W,self.h _ bias))
       sample _ h = self.sample _ from _ p(p _ h)
       return p _ h,sample _ h

    def h _ to _ v(self,h):
       p _ v = func.sigmoid(F.linear(h,self.W.t(),self.v _ bias))
       sample _ v = self.sample _ from _ p(p _ v)
       return p _ v,sample _ v

   def forward(self,v):
       pre _ h1,h1 = self.v _ to _ h(v)

       h _ = h1
       for _ in range(self.k):
           pre _ v _ ,v _ = self.h _ to _ v(h _ )
           pre _ h _ ,h _ = self.v _ to _ h(v _ )

       return v,v _
   def free _ energy(self,v):
       vbias _ term = v.mv(self.v _ bias)
       wx _ b = func.linear(v,self.W,self.h _ bias)
       hidden _ term = wx _ b.exp().add(1).log().sum(1)
       return (-hidden _ term - vbias _ term).mean()
rbm = RBM(k=1)
train _ op = optimize.SGD(rbm.parameters(),0.1)
for epoch in range(10):
   loss _ = []
       for _ , (data,target) in enumerate(train _ loader):
```

```
        data = Variable(data.view(-1,784))
        sample _ data = data.bernoulli()

        v,v1 = rbm(sample _ data)
        loss = rbm.free _ energy(v) - rbm.free _ energy(v1)
        loss _ .append(loss.data)
        train _ op.zero _ grad()
        loss.backward()
        train _ op.step()
    print("Training loss for {} epoch: {}".format(epoch,
np.mean(loss _ )))
def show _ adn _ save(file _ name,img):
    npimg = np.transpose(img.numpy(),(1,2,0))
    f = "./%s.png" % file _ name
    plt.imshow(npimg)
    plt.imsave(f,npimg)

show _ adn _ save("real",make _ grid(v.view(32,1,28,28).data))
show _ adn _ save("generate",make _ grid(v1.view(32,1,28,28).data))
```

When you execute that code, you should see something like what is shown in Figure 11.2.

After examining the previous code, the following quote may help further elucidate Boltzmann machines:

A Boltzmann machine is a network of symmetrically connected, neuron-like units that make stochastic decisions about whether to be on or off. Boltzmann machines have a simple learning algorithm (Hinton & Sejnowski, 1983) that allows them to discover interesting features that represent complex regularities in the training data. The learning algorithm is very slow in networks with many

```
Training loss for 0 epoch: -8.466686248779297
Training loss for 1 epoch: -6.686459064483643
Training loss for 2 epoch: -4.669025897979736
Training loss for 3 epoch: -3.197413921356201
Training loss for 4 epoch: -2.257789373397827
Training loss for 5 epoch: -1.6239577531814575
Training loss for 6 epoch: -1.0520764589309692
Training loss for 7 epoch: -0.7194262146949768
Training loss for 8 epoch: -0.4287303686141968
Training loss for 9 epoch: -0.27132782340049744
```

FIGURE 11.2 Boltzmann machine output.

layers of feature detectors, but it is fast in "restricted Boltzmann machines" that have a single layer of feature detectors. Many hidden layers can be learned efficiently by *composing* restricted Boltzmann machines, using the feature activations of one as the training data for the next.

Boltzmann machines are used to solve two quite different computational problems. For a *search* problem, the weights on the connections are fixed and are used to represent a cost function. The stochastic dynamics of a Boltzmann machine then allow it to sample binary state vectors that have low values of the cost function. For a *learning* problem, the Boltzmann machine is shown a set of binary data vectors and it must learn to generate these vectors with high probability. To do this, it must find weights on the connections so that, relative to other possible binary vectors, the data vectors have low values of the cost function. To solve a learning problem, Boltzmann machines make many small updates to their weights, and each update requires them to solve many different search problems.[10]

Restricted Boltzmann machines (RBM) are two layered building blocks that can learn a probability distribution over its set of input features. These are often used with feature learning, dimensionality reduction, classification, and regression. As the name suggests, RBMs are variations of Boltzmann machines, wherein the nodes/neurons form a bipartite graph. A bipartite graph (often called bigraph) is a graph whose vertices can be divided into two disjoint and independent sets (recall the introduction to graph theory in chapter 8). The result of this restricted Boltzmann machine is that it has fewer connections than a typical Boltzmann machine.

RADIAL BASIS FUNCTION NETWORK

As the name suggests, this is an artificial neural network that utilizes radial basis functions as the activation functions. This will, of course, require some explanation of radial basis functions. Radial basis functions (RBF), often symbolized by φ, have an output value that depends only on the distance between the input and some fixed point. That fixed point is typically either the origin or another fixed point c, called the center.

There are actually a variety of radial basis functions. Gaussian RBF is shown in equation 11.8.

$$\varphi(r) = e^{-(\varepsilon r)^2}$$
(eq. 11.8)

The symbol ε represents a shape parameter. Another version of the RBF is the inverse quadratic RBF shown in equation 11.9.

$$\varphi(r) = \frac{1}{1 + (\varepsilon r)^2}$$
(eq. 11.9)

Radial basis function networks usually have three layers: an input layer, a hidden layer, and an output layer. The hidden layer will have some type of RBF activation

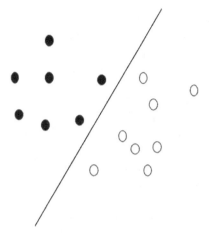

FIGURE 11.3 Linearly separable sets of points.

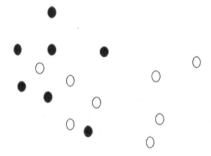

FIGURE 11.4 Non linearly separable sets of points.

function. What the hidden layer is doing is taking an input pattern that might not be linearly separable, and transforming it so that it is. If you are unfamiliar with the concept of linear separability, it comes from Euclidean geometry. Two sets of points are considered linearly separable if there exists at least one straight line that can separate all the points of one set from all the points of the other. This is shown in Figure 11.3.

Contrast that with two sets for which there is no straight line that can divide them. This is shown in Figure 11.4.

The radial basis functions help transform a pattern which may not be linearly separable into one that is. This process is based on Cover's theorem on the separability of patterns. That theorem states that a pattern that is transformed into a higher dimensional space with a nonlinear transformation has a greater likelihood of being linearly separable. This is why radial basis functions are nonlinear. This process also requires that the hidden layer have more nodes/neurons than the input layer. You can learn more about radial basis functions online.[11]

DEEP BELIEF NETWORK

A deep belief network is a type of deep neural network that is often described as a generative graphical model. Generative models provide a joint probability distribution over the input data and the labels for that data. DBNs can be used in both supervised and unsupervised learning scenarios.

DBNs were invented to solve the problem of deep layered networks functioning slowly, or becoming stuck in some local minima (usually due to improper parameter selection). DBNs have multiple layers with connections between the layers, but not between nodes in the same layer. DBNs are effective at learning to reconstruct the inputs after only a single set of examples. This allows the DBN to be utilized as an unsupervised feature detector. Then the DBN can be used via supervised learning for classification.

A deep belief network can be constructed using stacks of restricted Boltzmann machines (discussed earlier in this chapter). Each of the RBMs performs a nonlinear transformation on its input vectors. The output of that RBM is the input for the next RBM in the series.

SUMMARY

Neural networks are so common, with so many different variations, and this chapter provided an expanded coverage of artificial neural networks. Along with chapter 10, this chapter should provide you a working knowledge of neural networks, as well as experience working with several specific implementations of neural networks. It is important that you actually work through the examples provided. Reading the text can only provide a certain degree of knowledge. Experience working with the algorithms is key to truly learning machine learning for neuroscience.

EXERCISES

LAB 1: LSTM

The following is a sample LSTM Python script. It is not specifically focused on neuroscience, but should give you a basic understanding of LSTM.

```
import tensorflow as tf
import timeit
#note that the LSTM is built into
#tf.keras. You don't actually have to
# code much.
cell = tf.keras.layers.LSTMCell(10)
@tf.function
def fn(input, state):
return cell(input, state)
input = tf.zeros([10, 10])
state = [tf.zeros([10, 10])] * 2
```

```
cell(input, state)
fn(input, state)
dynamic _ graph _ time = timeit.timeit(lambda: cell(input, state),
number=100)
static _ graph _ time = timeit.timeit(lambda: fn(input, state),
number=100)
print('dynamic _ graph _ time:', dynamic _ graph _ time)
print('static _ graph _ time:', static _ graph _ time)
```

This does not display much on the screen, just two values:

```
dynamic_graph_time: 0.04790930007584393
static_graph_time: 0.03990430000703782
```

LAB 2: LSTM FOR NEUROSCIENCE

The dataset for this exercise can be found at www.kaggle.com/datasets/birdy654/eeg-brainwave-dataset-feeling-emotions. This dataset uses EEG to detect emotions. The dataset citation for this data is:

Bird, J. J., Manso, L. J., Ribiero, E. P., Ekart, A. and Faria, D. R. 2018. A study on mental state classification using eeg-based brain-machine interface. In *9th International Conference on Intelligent Systems.* IEEE.

Bird, J. J., Ekart, A., Buckingham, C. D. and Faria, D. R. 2019. Mental emotional sentiment classification with an eeg-based brain-machine interface. In *The International Conference on Digital Image and Signal Processing (DISP'19).* Springer.

```
#import your modules
import numpy as np
import pandas as pd
import seaborn as sns
import tensorflow as tf
import matplotlib.pyplot as plt
from sklearn.metrics import classification _ report
from sklearn.metrics import confusion _ matrix
from sklearn.metrics import accuracy _ score
from tensorflow.keras import Sequential
from tensorflow.keras.layers import Dense, Dropout
from tensorflow.keras.layers import Embedding
from tensorflow.keras.layers import LSTM
from sklearn.model _ selection import train _ test _ split
from tensorflow.keras.utils import to _ categorical
from sklearn.preprocessing import StandardScaler
#this assumes the csv data file is in the
```

```
#same directory as this script
data = pd.read_csv('emotions.csv')
#time series data we want to
#show in a graph
sampledata = data.loc[0, 'fft_0_b':'fft_749_b']
plt.figure(figsize=(10, 10)) #figure size
plt.plot(range(len(sampledata)), sampledata)
plt.title("Features fft_0_b to fft_749_b")
plt.show()
data['label'].value_counts()
labels = {'NEGATIVE': 0, 'NEUTRAL': 1, 'POSITIVE': 2}
data['label'] = data['label'].replace(labels)
X = data.drop('label', axis=1)
y = data['label'].copy()
scaler = StandardScaler()
scaler.fit(X)
X = scaler.transform(X)
y = to_categorical(y)
print (y)
print (type(y))
X_train, X_test, y_train, y_test = train_test_split(X, y,
train_size = 0.8, random_state = np.random)
X_train = np.reshape(X_train, (X_train.shape[0],1,X.shape[1]))
X_test = np.reshape(X_test, (X_test.shape[0],1,X.shape[1]))
tf.keras.backend.clear_session()
model = Sequential()
model.add(LSTM(64,    input_shape=(1,2548),activation="relu",ret
urn_sequences=True))
model.add(Dropout(0.2))
model.add(LSTM(32,activation="sigmoid"))
model.add(Dropout(0.2))
model.add(Dense(3, activation='sigmoid'))
#note the use of adam as the optimizer.
#you should experiment with different optimizers
model.compile(loss = 'categorical_crossentropy', optimizer =
"adam", metrics = ['accuracy'])
model.summary()
#10 epochs may not be enough, but it will execute faster
history = model.fit(X_train, y_train, epochs = 10, validation_
data= (X_test, y_test))
pred = model.predict(X_test)
predictionclasses= np.argmax(pred,axis=1)
expectedclasses = np.argmax(y_test,axis=1)
model_acc = accuracy_score(expectedclasses,predictionclasses)
print("Test Accuracy: {:.3f}%".format(model_acc * 100))
prediction = model.predict(X_test)
```

```
cm = confusion _ matrix(prediction.argmax(axis=1),y _ test.argmax
(axis=1))
print("Confusion Matrix\n",(cm))
plt.figure(figsize=(8, 8))
sns.heatmap(cm,    annot=True,    vmin=0,    fmt='g',    cbar=False,
cmap='Blues')
plt.xticks(np.arange(3) + 0.5, labels.keys())
plt.yticks(np.arange(3) + 0.5, labels.keys())
plt.xlabel("Predicted")
plt.ylabel("Actual")
plt.title("Confusion Matrix")
plt.show()
clr = classification _ report(prediction.argmax(axis=1), y _ test.
argmax(axis=1), target _ names=labels.keys())
print("Classification Report\n", clr)
```

Lab 3: Experiment with Activation Functions

We are going to take code from earlier in this book and experiment with things you learned in this chapter. The code is shown shortly. Note that the code uses softmax and relu6. You will experiment with changing those to SELU, SiLU, or the other activation functions you learned in this chapter. This will get you quite comfortable with activation functions.

```
import numpy as np
import pandas as pd # used for reading CSV data files
import os #needed for navigating file system
#The path is based on having downloaded the data to a
#subfolder of the directory this script is in and the
#folder was named 'braintumordataset'
for dirname, _ , filenames in os.walk('/BrainTumorDataSet'):
for filename in filenames:
print(os.path.join(dirname, filename))
#keras is used for our CNN
import keras
from keras.models import Sequential
from keras.layers import Conv2D,Flatten,Dense,MaxPooling2D,Dro
pout
from sklearn.metrics import accuracy _ score
import io
from PIL import Image
import tqdm
from sklearn.model _ selection import train _ test _ split
import cv2
from sklearn.utils import shuffle
import tensorflow as tf
```

```python
import matplotlib.pyplot as plt
xtrainarray = []
ytrainarray = []
imgsize = 150
labels = ['glioma _ tumor','meningioma _ tumor','no _
tumor','pituitary _ tumor']
for i in labels:
folderPath = os.path.join('BrainTumorDataSet/Training/',i)
for j in os.listdir(folderPath):
img = cv2.imread(os.path.join(folderPath,j))
img = cv2.resize(img,(imgsize,imgsize))
xtrainarray.append(img)
ytrainarray.append(i)
for i in labels:
folderPath = os.path.join('BrainTumorDataSet/Testing/',i)
for j in os.listdir(folderPath):
img = cv2.imread(os.path.join(folderPath,j))
img = cv2.resize(img,(imgsize,imgsize))
xtrainarray.append(img)
ytrainarray.append(i)
xtrainarray = np.array(xtrainarray)
ytrainarray = np.array(ytrainarray)
xtrainarray,ytrainarray = shuffle(xtrainarray,ytrainarray,ran
dom _ state=101)
xtrainarray.shape
xtrainarray,X _ test,ytrainarray,y _ test = train _ test _ split(xt
rainarray,ytrainarray,test _ size=0.1,random _ state=101)
ytrainarray _ new - []
for i in ytrainarray:
ytrainarray _ new.append(labels.index(i))
ytrainarray=ytrainarray _ new
ytrainarray = tf.keras.utils.to _ categorical(ytrainarray)
y _ test _ new = []
for i in y _ test:
y _ test _ new.append(labels.index(i))
y _ test=y _ test _ new
y _ test = tf.keras.utils.to _ categorical(y _ test)
# you should experiment with different activation functions
# Tanh, relu, relu6, gelu, etc.
model = Sequential()
model.add(Conv2D(32,(3,3),activation = 'relu6',input_shape=
(150,150,3)))
model.add(Conv2D(64,(3,3),activation= 'relu6'))
model.add(MaxPooling2D(2,2))
model.add(Dropout(0.3))
model.add(Conv2D(64,(3,3),activation='relu6'))
```

```
model.add(Conv2D(64,(3,3),activation='relu6'))
model.add(Dropout(0.3))
model.add(MaxPooling2D(2,2))
model.add(Dropout(0.3))
model.add(Conv2D(128,(3,3),activation='relu6'))
model.add(Conv2D(128,(3,3),activation='relu6'))
model.add(Conv2D(128,(3,3),activation='relu6'))
model.add(MaxPooling2D(2,2))
model.add(Dropout(0.3))
model.add(Conv2D(128,(3,3),activation='relu'))
model.add(Conv2D(256,(3,3),activation='relu'))
model.add(MaxPooling2D(2,2))
model.add(Dropout(0.3))
model.add(Flatten())
model.add(Dense(512,activation = 'relu'))
model.add(Dense(512,activation = 'relu'))
model.add(Dropout(0.3))
model.add(Dense(4,activation='softmax'))
model.summary()
model.compile(loss='categorical _ crossentropy',optimizer='Adam',
metrics=['accuracy'])
#5 epochs is just so this will run quickly. In real applications
you will often
# use more.
history = model.fit(xtrainarray,ytrainarray,epochs=5,validat
ion _ split=0.1)
acc = history.history['accuracy']
val _ acc = history.history['val _ accuracy']
epochs = range(len(acc))
fig = plt.figure(figsize=(14,7))
plt.plot(epochs,acc,'r',label="Training Accuracy")
plt.plot(epochs,val _ acc,'b',label="Validation Accuracy")
plt.legend(loc='upper left')
plt.show()
loss = history.history['loss']
val _ loss = history.history['val _ loss']
epochs = range(len(loss))
fig = plt.figure(figsize=(14,7))
plt.plot(epochs,loss,'r',label="Training loss")
plt.plot(epochs,val _ loss,'b',label="Validation loss")
plt.legend(loc='upper left')
plt.show()
img = cv2.imread('/BrainTumorDataSet/Training/pituitary _ tumor/p
(107).jpg')
img = cv2.resize(img,(150,150))
img _ array = np.array(img)
```

```
img _ array.shape
img _ array = img _ array.reshape(1,150,150,3)
img _ array.shape
from tensorflow.keras.preprocessing import image
img = image.load _ img('/BrainTumorDataSet/Training/pituitary _
tumor/p (107).jpg')
plt.imshow(img,interpolation='nearest')
plt.show()
a=model.predict(img _ array)
indices = a.argmax()

indices
```

NOTES

1. https://arxiv.org/pdf/1706.02515.pdf
2. https://arxiv.org/pdf/1702.03118.pdf
3. https://pytorch.org/docs/stable/generated/torch.nn.SiLU.html
4. www.nest-simulator.org/
5. https://github.com/kaizouman/tensorsandbox/blob/master/snn/leaky_integrate_fire.ipynb
6. https://cnrl.ut.ac.ir/SpykeTorch/doc/index.html
7. https://www.sciencedirect.com/science/article/pii/S2667096822000568#!
8. https://onlinelibrary.wiley.com/doi/full/10.1002/bimj.202000393
9. https://machinelearningmastery.com/gentle-introduction-long-short-term-memory-networks-experts/
10. www.scholarpedia.org/article/Boltzmann_machine
11. www.simplilearn.com/tutorials/machine-learning-tutorial/what-are-radial-basis-functions-neural-networks

12 K-Means Clustering

INTRODUCTION

K-means clustering was originally a method for vector quantization in signal processing. Vector quantization is a method of modeling probability density functions using the distribution of vectors. This technique has also been used for data compression. K-means clustering is an algorithm that seeks to find clusters in the dataset. K-means clustering is one of the easiest and unsupervised machine learning algorithms, which also makes it quite popular. The goal is rather simple. Take the input data and group similar data points together. This will aid in discovering patterns in the data. The term k-means was first used in 1967 by James MacQueen. There was a primitive version of the algorithm used by Bell Labs in 1957. The standard algorithm is often called naïve k-means. This algorithm is widely used in data mining.

There are some general clustering terms you will need to know as we move forward:

Davies–Bouldin score: The score is defined as the average similarity measure of each cluster with its most similar cluster, where similarity is the ratio of within-cluster distances to between-cluster distances. Thus, clusters that are farther apart and less dispersed will result in a better score.

Calinski–Harabasz: The score is defined as ratio of the sum of between-cluster dispersion and of within-cluster dispersion.

Silhouette score: The silhouette coefficient is calculated using the mean intra-cluster distance and the mean nearest-cluster distance for each sample.

Before we speak of specific clustering algorithms, there is a simple clustering script that uses patient data for patients with dementia. The dataset can be found at www.kaggle.com/code/dhwanimodi239/demo-dementia-classification/data?select=oasis_longitudinal.csv. We will use this same dataset and vary the code throughout this chapter.

```
import numpy as np
import pandas as pd
import seaborn as seab
import numpy as np
import matplotlib.pyplot as plt
df = pd.read _ csv('dementia/oasis _ longitudinal.csv')
#print first five rows of the dataset
df.head(5)
df.describe()
#determine number of subjects with dementia
```

DOI: 10.1201/9781003230588-15

```
seab.set _ style("whitegrid")
ex _ df = df.loc[df['Visit'] == 1]
palette=seab.color _ palette("terrain")
seab.countplot(x='Group', data=ex _ df,palette=palette)
print(palette[2])
#There are three groups so convert
ex _ df['Group'] = ex _ df['Group'].replace(['Converted'], ['With
Dimentia'])
df['Group'] = df['Group'].replace(['Converted'], ['With Dimentia'])
seab.countplot(x='Group', data=ex _ df,palette=palette)
# bar drawing function
def bar _ chart(feature):
    Dementia = ex _ df[ex _ df['Group']=='With Dementia'][feature].
value _ counts()
    NoDementia = ex _ df[ex _ df['Group']=='Without Dementia'][fea-
ture].value _ counts()
    df _ bar = pd.DataFrame([Dementia,NoDementia])
    df _ bar.index = ['Dementia','NoDementia']
    df _ bar.plot(kind='bar',stacked=True, figsize=(8,5))
    print(df _ bar)


#plot out data by age
plt.figure(figsize=(10,5))
seab.violinplot(x='CDR', y='Age', data=df)
plt.title('Violin plot of Age by CDR',fontsize=14)
plt.xlabel('Clinical Dementia Rating (CDR)',fontsize=13)
plt.ylabel('Age',fontsize=13)
plt.show()
```

When executed, you will see two graphs displayed. The first graph is shown in Figure 12.1.

In previous chapters, you have seen other types of graphs used. You should experiment with some of these graph types. The violin graph is just exemplary.

K-MEANS CLUSTERING

The essence of the algorithm begins by defining the number of centroids you want to find in the data set. This is the target number, k. A centroid is just the center of a cluster. Essentially you are defining how many clusters you want produced from the dataset. At this point, you won't know what data will be in a cluster, nor how it is clustered. You will have simply defined how many clusters you are trying to find. Each data point is then allocated to the centroid that it has the closest match to. The algorithm starts with the first group of randomly selected centroids (you choose the number, but the values are randomly chosen). Then there is an iterative set of calculations to determine the optimum positions (values) for the centroids. Once those

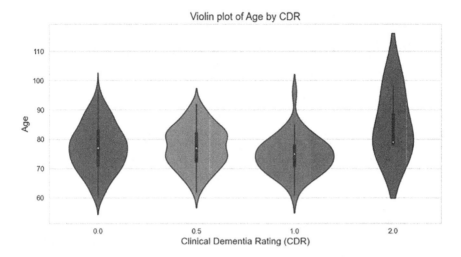

FIGURE 12.1 Violin graph.

have been stabilized, then the algorithm can stop. The scikit package is often used for k-means clustering. With that package you have the following parameters:

max_iter: the maximum number of iterations. The default is 300.

Init: This is the initialization of the centroids.

n_init: The number of times the algorithm will execute using a different seed each time.

Algorithm: There are many variations of k-means clustering you can use. The default is the Lloyd or naïve k-means.

tol: The tolerance regarding the Frobenius norm of the difference in the cluster centers of two consecutive iterations. This is used to determine if convergence has occurred.

cluster_centers_: This is an array of coordinates for cluster centers.

labels_: This is an array of labels for each point.

inertia_: This is a floating point value that has the sum of squared distances of samples with their closest cluster center.

There are various initialization methods. Two of the most common are the random partition and the Forgy method. The random partition method, as the name suggests, begins by randomly assigning a cluster to each data point, and updating from there. This computes an initial mean to the centroid of the cluster and its randomly assigned points. The Forgy method randomly chooses k observations from the dataset and simply uses these as the initial means.

The following code is a very basic k-means clustering script. It is simply meant to get you comfortable with the k-means process.

```
import numpy as np
class kmeans:
```

```
""""" The k-Means algorithm"""""
 def _ _ init _ _ (self,k,data):
      #set number of datapoints or input vectors
      self.nData = np.shape(data)[0]
      #set number of dimensions or features
      self.nDim = np.shape(data)[1]
      #set number of clusters
      self.k = k
 def kmeanstrain(self,data,maxIterations=10):
      # Find the minimum and maximum values for each feature
      minima = data.min(axis=0)#gives 1 X nDim matrix
      maxima = data.max(axis=0)#gives 1 X nDim matrix
      # Pick the center locations randomly
      #each row contains a cluster centre
      #here random matrix with 0-1 is generated multiplied by
(max-min) for each feature
      # then min of feature is added. * is for element wise
multiplication of matrix
      self.centres = np.random.rand(self.k,self.nDim)*(maxima-
minima)+minima
      oldCentres = np.random.rand(self.k,self.nDim)*(maxima-
minima)+minima
      count = 0#tracks the no of interation
       #run to maximum interations passed in or unless the cen-
ters doesn't move
      while np.sum(np.sum(oldCentres-self.centres))!= 0 and count
<maxIterations:
            oldCentres = self.centres.copy()
            count += 1
            # Compute euclidean distances from each centre
            #each row is for the cluster center and column is
distance to the centre
            distances = np.ones((1,self.nData))*np.sum((data-self.
centres[0,:])**2,axis=1)
            for j in range(self.k-1):
            distances = np.append(distances,np.ones((1,self.nData))
*np.sum((data-self.centres[j+1,:])**2,axis=1),axis=0)
                # Identify the closest cluster
                #argmin returns the index of row
                #cluster is 1 X nData matrix where each cell is index
for cluster centre
                cluster = distances.argmin(axis=0)
                cluster = np.transpose(cluster*np.ones((1,self.nData)))
                # Update the cluster centres
                for j in range(self.k):
                    #include or exclude element to the current cluster
with 1 or 0
```

```
                thisCluster = np.where(cluster==j,1,0)
                if sum(thisCluster)>0:#if any datapoint belong to
cluster update the center by taking mean
                    self.centres[j,:] = np.sum(data*thisCluster,
axis=0)/np.sum(thisCluster)
        return self.centres
```

The following code sample utilizes k-means clustering with MRI/Alzheimer's data from Kaggle. This code depends on the data set from Kaggle. You can download that from https://www.kaggle.com/datasets/jboysen/mri-and-alzheimers. It should also be noted that this is a longer script than we have been using, but it also provides a wide range of visual outputs.

```
import numpy as np
import pandas as pd
import matplotlib.pyplot as plot
import seaborn as seab
import os
import warnings
# KMeans Clustering
from sklearn.cluster import KMeans
from sklearn.preprocessing import StandardScaler
from sklearn.metrics import mean_squared_error
# it is likely you won't have seaborn
# just use pip install seaborn
import seaborn as seab
from sklearn.model_selection import train_test_split
warnings.filterwarnings("ignore")
#read in the data, make sure you have downloaded it from
# https://www.kaggle.com/code/pragyamukherjee/
alzheimer-s-k-means/data
data=pd.read_csv("alzheimerdata/oasis_cross-sectional.csv")
data.head()
data.Hand.unique()
data.drop(["ID","Hand","Delay"],axis=1,inplace=True)
data.columns=["gender","age","education","soc_eco_st","mini_
mental_state_exam","clinical_dementia_rating",

        "estimated_total_intracranial_volume","normalize_
whole_brain_volume","atlas_scaling_factor"]
data.gender = [1 if each == "F" else 0 for each in data.gender]
data.info()
data.isnull().sum()
data.describe()
def impute_median(series):
    return series.fillna(series.median())
data.education =data["education"].transform(impute_median)
```

```
data.soc_eco_st =data["soc_eco_st"].transform(impute_median)
data.mini_mental_state_exam =data["mini_mental_state_
exam"].transform(impute_median)
data.clinical_dementia_rating =data["clinical_dementia_
rating"].transform(impute_median)
#visualize the correlation
plot.figure(figsize=(15,10))
seab.heatmap(data.iloc[:,0:10].corr(), annot=True,fmt=".0%")
plot.show()
kmeans = KMeans(
init="random",
n_clusters=2,
n_init=10,
max_iter=300,
random_state=42
)
kmeans.fit(data.loc[:,['age','normalize_whole_brain_volume']])
pred = kmeans.predict(data.loc[:,['age','normalize_whole_brain_
volume']])
plot.scatter(data= data,x ='age',y = 'normalize_whole_brain_
volume',c = pred, cmap='viridis')
plot.xlabel('age')
plot.ylabel('clinical_dementia_rating')
plot.show()
kmeans_kwargs = {"init": "random","n_init": 10,"max_iter": 300,
"random_state": 42,}
sse = []
for k in range(1, 6):
    kmeans = KMeans(n_clusters=k, **kmeans_kwargs)

    kmeans.fit(data.loc[:,['age','normalize_whole_brain_volume']])
    sse.append(kmeans.inertia_)

plot.style.use("fivethirtyeight")
plot.plot(range(1, 6), sse)
plot.xticks(range(1, 6))
plot.xlabel("No. of Clusters")
plot.ylabel("SSE")
# Sum of Squares (SSE) is the sum of the squared differences
# between each observation and that group's mean.
plot.show()
```

When you execute this script, you will have multiple outputs. First the command line output, which is shown in Figure 12.2.

But in addition to this output, several graphs are also output. These are shown in Figures 12.3 through 12.5. Note that subsequent images only show after you have closed the previous image.

```
E:\Projects\publishing\Machine Learning For Neuroscience>python kmeansexample.py
<class 'pandas.core.frame.DataFrame'>
RangeIndex: 436 entries, 0 to 435
Data columns (total 9 columns):
 #   Column                              Non-Null Count  Dtype
---  ------                              --------------  -----
 0   gender                              436 non-null    int64
 1   age                                 436 non-null    int64
 2   education                           235 non-null    float64
 3   soc_eco_st                          216 non-null    float64
 4   mini_mental_state_exam              235 non-null    float64
 5   clinical_dementia_rating            235 non-null    float64
 6   estimated_total_intracranial_volume 436 non-null    int64
 7   normalize_whole_brain_volume        436 non-null    float64
 8   atlas_scaling_factor                436 non-null    float64
dtypes: float64(6), int64(3)
memory usage: 30.8 KB
```

FIGURE 12.2 K-means Alzheimer's output.

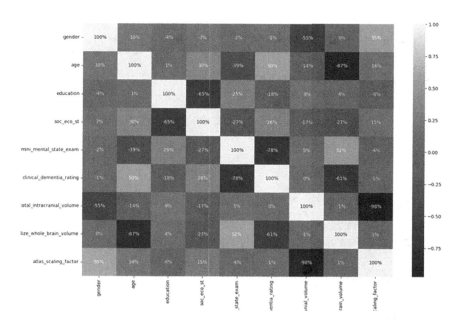

FIGURE 12.3 K-means Alzheimer's chart.

Again, this is a long example script. But as can be seen from the data presentation, it contains a number of items that are well worth studying. It is strongly suggested that you take the time to carefully analyze this script. You can also find similar scripts on the Kaggle website, in the same location you downloaded the dataset.

Rather than use sckit, you could also use the sklearn library.[1] In that library is sklearn.cluster.kMeans, with the following parameters:

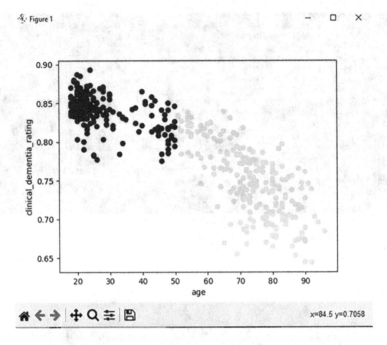

FIGURE 12.4   K-means Alzheimer's scatter plot.

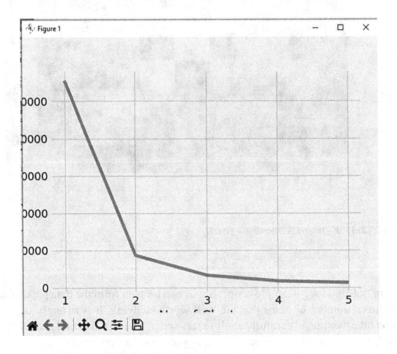

FIGURE 12.5   K-means Alzheimer's graph.

n_clusters: The number of clusters/centroids to form. The default is 8.

n_init: The number of times the algorithm will run with different centroid seeds. The default is 10.

Randomstate: The random-number generation for initializing centroids. The default is none.

Algorithm: Which version of k-means clustering to use. The possibilities are "Lloyd" (also the default), "elkan," "auto," and "full."

As you can see, you have a wide variety of options with k-means clustering. You can use scikit or sklearn, and both have similar parameters you can work with. As with any algorithm, the choice of parameters will determine the efficacy and efficiency of your machine learning script. However, there are no definite rules for the choice of these parameters. One normally spends some time experimenting with different parameters, seeking an ideal set of values.

K-MEANS++

This algorithm is much like k-means, but focuses on the choice of the initial seed values. The goal of k-means is to find cluster centers that minimize the variance within the class. This is often determined by the sum of squared distances from each data point to its cluster center. K-means++ is directed towards finding the initial cluster centers, thus improving the efficacy of k-means. The steps of this algorithm are given here:

1. Select one center uniformly at random among the data points.
2. For each data point x not yet selected, calculate the distance between x and the nearest center (Dx) that has already been selected.
3. Select one new data point at random as a new center, using a weighted probability distribution where a point x is chosen with probability proportional to $D(x)^2$.
4. Repeat steps 2 and 3 until k centers have been chosen.
5. Now that the initial centers have been chosen, proceed using standard k-means clustering.

As you can see, this process is about selecting better centroids for k-means clustering. Many libraries allow you to use k-means++ as the k-means algorithm. The following code implements a basic k-means++.

```
#importing required libraries
import numpy as nump
import pandas as pd
import matplotlib.pyplot as plt
import sys
# first we create some data to use
# our algorithm on. The numbers chosen here
# are randomly selected. Choose any numbers
```

```
# you wish
firstmean = nump.array([0.0, 0.0])
cov _ 01 = nump.array([[2, 0.3], [0.3, 2]])
firstdistance   =   nump.random.multivariate _ normal(firstmean,
cov _ 01, 100)
secondmean = nump.array([4.0, 5.0])
cov_ 02 = nump.array([[2.5, 0.3], [0.3, 1]])
seconddistance = nump.random.multivariate _ normal(secondmean,
cov _ 02, 100)
thirdmean = nump.array([4.0, -5.0])
cov _ 03 = nump.array([[1.5, 0.5], [0.5, 1,5]])
thirddistance  =  nump.random.multivariate _ normal(thirdmean,
cov _ 01, 100)
fourthmean= nump.array([2.0, -9.0])
cov _ 04 = nump.array([[1.2, 0.3], [0.5, 1,2]])
fourthdistance = nump.random.multivariate _ normal(fourthmean,
cov _ 01, 100)
data  =  nump.vstack((firstdistance,  seconddistance,  thirddis-
tance, fourthdistance))
nump.random.shuffle(data)
# this function will plot initial centroids
def plotcentroids(data, centroids):
    plt.scatter(data[:, 0], data[:, 1], marker = '.',
        color = 'red', label = 'data points')
    plt.scatter(centroids[:-1, 0], centroids[:-1, 1],
        color = 'blue', label = 'previously selected centroids')
    plt.scatter(centroids[-1, 0], centroids[-1, 1],
        color = 'green', label = 'next centroid')
    plt.title('Select centroid number % d'%(centroids.shape[0]))
    plt.legend()
    plt.xlim(-5, 12)
    plt.ylim(-10, 15)
    plt.show()
# function to compute euclidean distance
def distance(p1, p2):
 return nump.sum((p1 - p2)**2)
# initialization algorithm
def initialize(data, k):
    '''
    initialized the centroids for K-means++
        inumputs:
        data - numpy array of data points having shape (200, 2)
        k - number of clusters
    '''
    ## initialize the centroids list and add
    ## a randomly selected data point to the list
```

```
    centroids = []

centroids.append(data[nump.random.randint(data.shape[0]), :])
    plotcentroids(data, nump.array(centroids))
    ## compute remaining k - 1 centroids
    for c_id in range(k - 1):
    ## initialize a list to store distances of data
    ## points from nearest centroid
        dist = []
        for i in range(data.shape[0]):
            point = data[i, :]
            d = sys.maxsize
            ## compute distance of 'point' from each of the
previously
            ## selected centroid and store the minimum distance
            for j in range(len(centroids)):
                temp_dist = distance(point, centroids[j])
                d = min(d, temp_dist)
            dist.append(d)
        ## select data point with maximum distance as our
next centroid
        dist = nump.array(dist)
        next_centroid = data[nump.argmax(dist), :]
        centroids.append(next_centroid)
        dist = []
        plotcentroids(data, nump.array(centroids))
    return centroids
# call the initialize function to get the centroids
centroids = initialize(data, k = 4)
```

When you execute this code, you will see four graphs showing the progression of the algorithm. These graphs are shown in Figures 12.6 to 12.9.

The goal of this code is to acclimate you to the k-means++ algorithm. You should review this code along with previously presented k-means code, and ensure you understand the differences. In the exercises at the end of the chapter you will use this code with neuroscience data.

K-MEDIANS CLUSTERING

As the name suggests, the k-medians clustering is much like the k-means clustering, except that it utilizes the statistical median, rather than the mean. With this variation, the median is computed using the Manhattan-distance formula. This is sometimes call taxicab geometry. Rather than use the Euclidean distance between two points, this method considers grid layouts, such as in city blocks. Put more formally, the Manhattan distance, d1 between two vectors p, q in an n-dimensional space with a fixed Cartesian coordinate system is the sum of the lengths of the projections of the

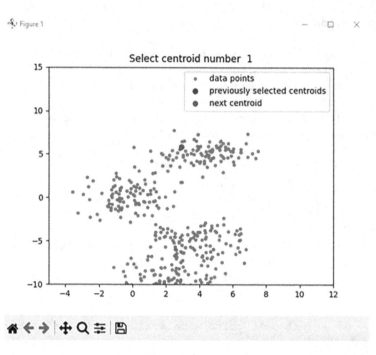

FIGURE 12.6 K-means++ centroid one.

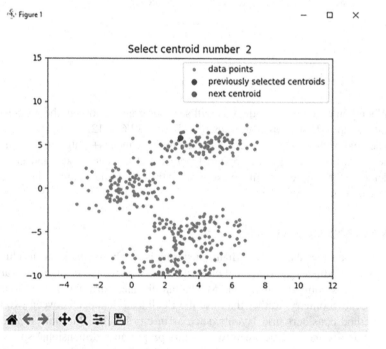

FIGURE 12.7 K-means++ centroid two.

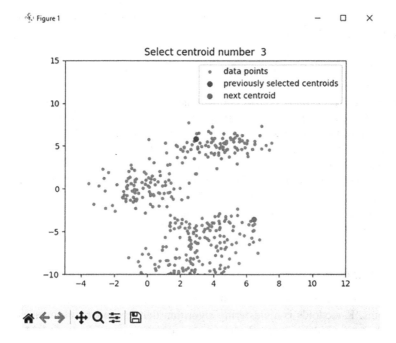

FIGURE 12.8   K-means++ centroid three.

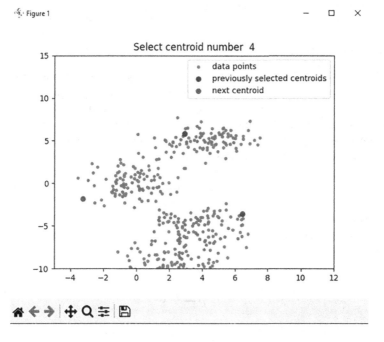

FIGURE 12.9   K-means++ centroid four.

line segment between the points onto the coordinate axes. The formal definition of the Manhattan distance is given in equation 12.1.

$$dT(p,q) = \|p - q\| \; T = \sum\nolimits_{i=1}^{n} |p_i - q_i|$$ (eq. 12.1)

In equation 12.1, d_T is the taxicab distance. The distance is between two vectors, $p = (p_1, p_2, \ldots p_n)$ and $q = (q_1, q_2, \ldots q_n)$. While in our application, this is being used with machine learning, the algorithm itself predates machine learning considerably. This metric was first used in 1757 by Roger Boscovich to perform regression analysis.

K-medians clustering tends to minimize error over clusters by using the median rather than the mean for each cluster. This is particularly useful if your dataset has outliers. K-medians will essentially ignore outliers, whereas k-means will skew your cluster centroids with the outliers' value.

K-MEDOIDS

Sometimes k-medians is confused with k-medoids, however they are quite different algorithms. K-medoids is a clustering algorithm that is similar to k-means. Rather than select centroids then match data points to those centroids, k-medoids selects some of the actual data points as centers. This is where the name of this algorithm originates. A medoid is a representative object of a dataset whose sum of dissimilarities to all the other objects in the dataset is minimal. There are variations of k-medoids such as Partitioning Around Medoids (PAM).

RANDOM FOREST

The Random Forest algorithm is not as common as some of the other algorithms discussed in this chapter. Random Forest begins with bootstrap aggregating (sometimes referred to as bagging). The process of bootstrap aggregating involves a training algorithm that takes the training set and repeatedly selects a random sample with replacement of the training set. The training set is described in equation 12.2, where X is the training set and Y represents the responses.

$$X = x_1, \ldots, X_n; \; Y = y_1, \ldots, y_n$$ (eq 12.2)

The bootstrap aggregating is executed B times. After training, unseen samples of x' can be predicted from the sum of the individual regression trees on x'. This is shown in equation 12.3.

$$\hat{f} = \frac{1}{B} \sum_{b=1}^{B} f_b(x')$$ (eq. 12.3)

This technique reduces variance in the model without increasing bias. Some researchers consider Random Forest to be related to k-nearest neighbor in that both are weighted neighborhood approaches. The concept of the forest is simply that a

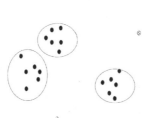

FIGURE 12.10   DBSCAN.

large number of uncorrelated models (trees) operating as a group will perform better than any of the individual constituent models.

DBSCAN

This algorithm was given a passing mention in chapter 9. It is a common clustering algorithm, and thus related to k-means clustering. DBSCAN is an acronym for *density-based spatial clustering of applications with noise*. The term density-based requires some explanation. Density-based clustering involves grouping together points that are closely packed together, and those that lie in very low density regions are considered noise. This is illustrated in Figure 12.10.

In Figure 12.10, those data points that have others nearby are clustered into groups. These are represented by solid black data points. Those data points that are far from any other data points are treated as noise. Those are represented by striped circles. The algorithm essentially has the following criteria for clustering:

- A point p is a core point if there are at least some minimum number of other points within distance ε of point p. You decide what the minimum number of points is and the distance.
- A point q is considered directly reachable from point p, if point q is within distance ε of point p.
- A point q is reachable (note: not directly reachable, just reachable) from p if there is some path p1, . . . , pn, and pn = q where each pi + 1 is directly reachable from pi. In other words, you can trace a series of points from p to q wherein each point is directly reachable from the previous and subsequent points.
- Any points that are not reachable are considered noise.

Thus, to use DBSCAN, you need to set parameters, the minimum number of points in an area to consider it dense (minimum points) and the distance ε that constitutes nearby. The scikit-learn library has a DBSCAN sklearn.cluster.DBSCAN, you just set the parameters and you are using DBSCAN.

SUMMARY

This chapter has focused on various clustering functions, beginning with k-means clustering. Clustering functions allow one to detect patterns in the data, even when a specific goal has not been established. This makes clustering algorithms of particular importance for processes such as datamining.

EXERCISES

EXERCISE 1: K-MEANS WITH ALZHEIMER'S DATA

For this exercise you will use code we examined earlier in this chapter. Your goal is to first make it work, then to experiment. Try changing the k-means parameters to see what effect that has:

```
init="random",
n _ clusters=2,
n _ init=10,
max _ iter=300,
random _ state=42
```

```
import numpy as np
import pandas as pd
import matplotlib.pyplot as plot
import seaborn as seab
import os
import warnings
# K-Means Clustering
from sklearn.cluster import KMeans
from sklearn.preprocessing import StandardScaler
from sklearn.metrics import mean _ squared _ error
# it is likely you won't have seaborn
# just use pip install seaborn
import seaborn as seab
from sklearn.model _ selection import train _ test _ split
warnings.filterwarnings("ignore")
#read in the data, make sure you have downloaded it from
# https://www.kaggle.com/datasets/jboysen/mri-and-alzheimers
data=pd.read _ csv("alzheimerdata/oasis _ cross-sectional.csv")
data.head()
data.Hand.unique()
data.drop(["ID","Hand","Delay"],axis=1,inplace=True)
data.columns=["gender","age","education","soc _ eco _ st","mini _
mental _ state _ exam","clinical _ dementia _ rating",
"estimated _ total _ intracranial _ volume","normalize _ whole _
brain _ volume","atlas _ scaling _ factor"]
data.gender = [1 if each == "F" else 0 for each in data.gender]
data.info()
data.isnull().sum()
data.describe()
```

```
def impute _ median(series):
return series.fillna(series.median())
data.education =data["education"].transform(impute _ median)
data.soc _ eco _ st =data["soc _ eco _ st"].
transform(impute _ median)
data.mini _ mental _ state _ exam =data["mini _ mental _ state _
exam"].transform(impute _ median)
data.clinical _ dementia _ rating =data["clinical _ dementia _
rating"].transform(impute _ median)
#visualize the correlation
plot.figure(figsize=(15,10))
seab.heatmap(data.iloc[:,0:10].corr(), annot=True,fmt=".0%")
plot.show()
kmeans = KMeans(
init="random",
n _ clusters=2,
n _ init=10,
max _ iter=300,
random _ state=42
)
kmeans.fit(data.loc[:,['age','normalize _ whole _ brain _ volume']])
pred = kmeans.predict(data.
loc[:,['age','normalize _ whole _ brain _ volume']])
plot.scatter(data= data,x ='age',y = 'normalize _ whole _ brain _
volume',c = pred, cmap='viridis')
plot.xlabel('age')
plot.ylabel('clinical _ dementia _ rating')
plot.show()
kmeans _ kwargs = {"init": "random","n _ init": 10,"max _ iter":
300,"random _ state": 42,}
sse = []
for k in range(1, 6):
kmeans = KMeans(n _ clusters=k, **kmeans _ kwargs)
kmeans.fit(data.loc[:,['age','normalize _ whole _ brain _ volume']])
sse.append(kmeans.inertia _ )
plot.style.use("fivethirtyeight")
plot.plot(range(1, 6), sse)
plot.xticks(range(1, 6))
plot.xlabel("No. of Clusters")
plot.ylabel("SSE")
# Sum of Squares (SSE) is the sum of the squared differences
# between each observation and that group's mean.
1.  plot.show()
```

Exercise 2: K-Means++ with Neurological Data

For this exercise you will begin with the k-means++ code given earlier in this chapter. However, you will modify it to import neuroscience data. This will be a challenging lab and is intended for those readers that are easily mastering the previous labs (both in this chapter as well as chapters 10 and 11). One approach is to import the Alzheimer's data from exercise 1, then combine that with the k-means++ clustering shown earlier in this chapter. This will require some modification to the code, you won't be able to just cut and paste everything.

NOTE

1. https://scikit-learn.org/stable/modules/generated/sklearn.cluster.KMeans.html

13 K-Nearest Neighbors

INTRODUCTION

Classification normally requires partitioning samples into training and testing categories. Let $\mathbf{x}i$ be a training sample and $\mathbf{x}$ be a test sample, and let ω be the class of a training sample and $\omega^\wedge$ be the predicted class for a test sample $(\omega,\omega^\wedge = 1,2, \ldots ,\Omega)$. Here, Ω is the total number of classes. This may seem like an overly mathematical representation of the concept, but is actually simpler than it appears. The goal is to use training algorithms to find classes in the data, then match test data to the appropriate classes.

The k-nearest neighbors algorithm was first developed long before machine learning. It was created by Evelyn Fix and Joseph Hodges in 1951 for use in statistics. This algorithm can be used for classification or regression.

When used for classification, the concept is relatively simple: take the input samples and determine what the closest classification is, and classify the input data. The algorithm uses training examples that are vectors in some multidimensional space (recall chapter 1's discussion of vectors). The algorithm then determines which classification is closest to the input data.

When working with continuous variables, the most commonly used algorithm for determining the closest classification is Euclidean distance. Euclidean distance is the length of a line segment between two points. There are many variations of this formula. The first, shown in equation 13.1, is for data that has two dimensions. There is a point p and a point q, with coordinates. The distance between p and q is shown in equation 13.1.

$$d(p,q) = \sqrt{\left(q_1 - p_1\right)^2 + \left(q_2 - p_2\right)^2} \qquad \text{(eq. 13.1)}$$

If the data has additional dimensions, then that can be extended as shown in equation 13.2 for three dimensions.

$$d(p,q) = \sqrt{\left(p_1 - q_1\right)^2 + \left(p_2 - q_2\right)^2 + \left(p_3 - q_3\right)^2} \qquad \text{(eq. 13.2)}$$

It is also possible to compute distance using polar coordinates. The polar coordinates of a given point p are (r, θ). This gives the formula shown in equation 13.3.

$$d(p,q) = \sqrt{r_p^2 + r_q^2 - 2r_p r_q \left(\theta_p - \theta q\right)} \qquad \text{(eq. 13.3)}$$

If discrete variables are used then, one can use Hamming distance or a similar metric. Discrete variables are often used for applications such as text classification. Hamming distance calculates the difference between two strings. The Hamming distance

DOI: 10.1201/9781003230588-16

is, essentially, the number of characters that are different between two strings. This can be expressed mathematically as h(x, y). The concept of Hamming distance was developed by Richard Hamming who first described the concept in his paper "Error detecting and error correcting codes."

A basic Python function for computing Hamming distance is shown in the following code.

```
def hamming _ distance(string1, string2):
dist _ counter = 0
for n in range(len(string1)):
if string1[n]!= string2[n]:
dist _ counter += 1
return dist _ counter,
```

Another similar metric is the Levenshtein distance. This is a measurement of the number of single-character edits required to change one word into another. Edits can include substitutions (as with Hamming distance) but can also include insertions and deletions. The Levenshtein distance was first described by Vladimir Levenshtein in 1965.

Regardless of the types of data being used or the distance algorithm selected, the k is how many nearest neighbors are used to decide if this current point is in that classification or not. For example, if k is set to one, then data points are classified simply according to one nearest neighbor. If k is set to five, then the five nearest neighbors are used to determine the classification of the current data point. It is also common to assign weights to the neighbors so that the closest neighbors carry the most weight. If, for example, you have k = 5, that means the 5 nearest neighbors will determine classification. But what if those 5 are not all in the same class? By assigning weights, you ensure that the closest neighbors will be the determining factor. Consider Figure 13.1.

In Figure 13.1, the black dots are data points that are already in groups (group 1 and 2). The d represents a new data point. The dashed box surrounds the five nearest neighbors. In this example, the five nearest neighbors are in two different groups. However, it is easy to see two of the black dots are much closer to the data point d than are the others. This should give them more weight. This would lead to the data point d being assigned group 1, rather than group 2.

Another issue with KNN is the dimensionality. Essentially, KNN performs best with a lower number of features. More features require more data, and with KNN that can lead to the issue of overfitting.[1] Overfitting occurs because when there is a high number of features, noise ends up skewing classifications.

EXAMINING KNN

The KNN algorithm is a fairly common classification method and is thus worth studying. The following code is a basic example of using KNN with the iris database, which we have used previously in this book.

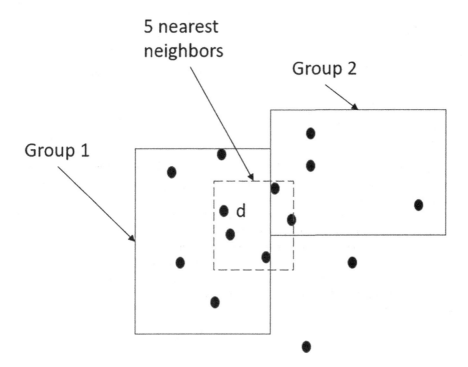

FIGURE 13.1 KNN weighted data points.

```
#import necessary modules
from sklearn.neighbors import KNeighborsClassifier
from sklearn.model_selection import train_test_split
from sklearn.datasets import load_iris
import numpy as np
import matplotlib.pyplot as plt
iris_data = load_iris()
# Create feature and target arrays
X = iris_data.data
y = iris_data.target
# Split into training and test data
X_train, X_test, y_train, y_test = train_test_split(
  X, y, test_size = 0.2, random_state=30)
neighbors = np.arange(1, 10)
train_accuracy = np.empty(len(neighbors))
test_accuracy = np.empty(len(neighbors))
# Loop over k-values and compute accuracy
for i, k in enumerate(neighbors):
  knn = KNeighborsClassifier(n_neighbors=k)
  knn.fit(X_train, y_train)
  train_accuracy[i] = knn.score(X_train, y_train)
```

```
      test_accuracy[i] = knn.score(X_test, y_test)
# Generate plot
plt.plot(neighbors, test_accuracy, label = 'Testing Accuracy')
plt.plot(neighbors, train_accuracy, label = 'Training Accuracy')
plt.legend()
plt.xlabel('n_neighbors')
plt.ylabel('Accuracy')
plt.show()
```

When you execute that code, the output will be similar to what is shown in Figure 13.2. Note, you can also experiment with changing the values for n_neighbors and random_state.

You can see the effect of KNN in the preceding example. You should take the time to experiment with that example, even changing parameters, until it seems very clear to you. It is also recommended that you vary your choice of k and determine which value of k is most effective for a given problem. The choice of k is critical. Generally, larger k values reduce noise in the data but also make class boundaries less distinct. It is also important to know that k-nearest neighbors is built into scikit-learn. It is sklearn.neighbors.

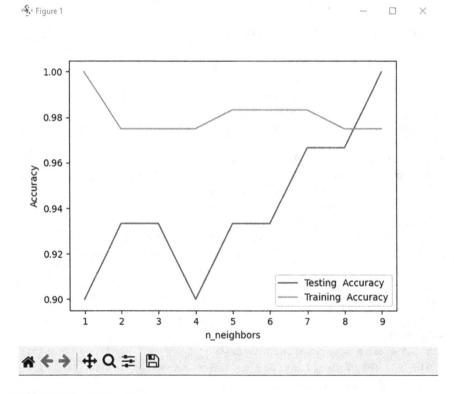

FIGURE 13.2 KNN output.

DIMENSIONALITY REDUCTION

KNN depends on dimensionality reduction. There are several algorithms that can be used for this purpose. Principle component analysis, which was discussed in chapter 9, is one of the most common. A brief refresher on PCA is given here. PCA is essentially a statistical technique for reducing the dimensionality of a dataset. The principal components for a set of points in a real coordinate space are a series of p unit vectors where the ith vector is the direction of a line that best fits the data.

Another option is linear discriminant analysis. Linear discriminant analysis (LDA) is closely related to the ANOVA test that was discussed in chapter 2. A brief refresher on ANOVA is given here. ANOVA, or analysis of variation, is actually a family of statistical tests. We will focus on just one. The one-way analysis of variance (ANOVA) is used to determine whether there are any statistically significant differences between the means of two or more independent groups. The one-way ANOVA compares the means between the groups and determines if the means are statistically different from each other.

LDA is a method to find a linear combination of features that either characterizes or separates two or more classes. The combination resulting from the process can be used as a linear classifier or for dimensionality reduction. Consider a machine learning situation where you have a number of classes, each with multiple features. If you use, for example, just one feature to classify data, you will have a great deal of overlap. This problem is particularly important when applying machine learning to neuroscience. Recall in chapter 8 when various neurological disorders were discussed. Many had overlapping features.

LDA uses both the X and Y axes to create a new axis, then projects the data onto that new axis to maximize the separation of the data. LDA will maximize the distance between two classes, and minimize the variation within each class. The details of LDA depend heavily on the linear algebra in chapter 1 and the statistics in chapter 2. For our purposes, we don't need to delve into those details. As you probably surmised by now, the details of the LDA algorithm are handled by Python libraries.

While Euclidean distance is the most common distance algorithm you will see in machine learning, there are other distance algorithms. In a previous chapter you were introduced to Manhattan distance. Another distance formula is called Mahalanobis distance and is named after Presanta Mahalanobis, a statistician who introduced this method in 1936.[2] Mahalanobis distance is the distance between a point P and a distribution D. Or put another way, it is the distance between two points, in multivariate space. This method is often used to find multivariable outliers.

VISUALIZE KNN

This section is simply to assist you in getting comfortable with KNN. The data source is not important; however, you can download the one used in this example from. The following code displays the data as the algorithm is executed, thus letting you see what is happening inside the KNN algorithm.

```
import numpy as np # linear algebra
import pandas as pd # data processing, CSV file I/O (e.g.,
pd.read_csv)
```

```
import sklearn
import matplotlib.pyplot as plt
import seaborn as sns
from sklearn.utils import shuffle
from sklearn import neighbors, metrics
from IPython.display import display
#import data
df = pd.read_csv("concertriccir2.csv")
#first shuffle data
#then display it
df = shuffle(df, random_state = 15)
df = df.reset_index()
X = df.iloc[:, 1:3]
display(X)
y = df.iloc[:, 3:]
display(y)
#now plot the data, before
#working with it
sns.scatterplot(data = df.iloc[:400,:], x = 'X', y = 'Y',
hue = "class")
#now set up data sets
X_train = X.iloc[:400,:]
X_val = X.iloc[400:450,:]
X_test = X.iloc[450:,:]
y_train = y.iloc[:400,:]
y_val = y.iloc[400:450,:]
y_test = y.iloc[450:,:]
display(X_test)
#now we can use KNN
model = sklearn.neighbors.KNeighborsClassifier(5)
model.fit(X_train, y_train)
y_pred = model.predict(X_train)
print(sklearn.metrics.accuracy_score(y_train, y_pred))
y_pred_val = model.predict(X_val)
print(sklearn.metrics.accuracy_score(y_val, y_pred_val))
plt.figure(figsize=(20,10))
sns.scatterplot(x = X_train.loc[:, "X"], y = X_train.loc[:,
"Y"], hue = y_train.loc[:, "class"], style = y_pred)
sns.scatterplot(x = X_val.loc[:, "X"], y = X_val.loc[:, "Y"],
hue = y_val.loc[:, "class"], style = y_pred_val)
k_values = []
accuracies = []
for i in range(1, 15):
    model = sklearn.neighbors.KNeighborsClassifier(i)
    model.fit(X_train, y_train)
    y_pred = model.predict(X_train)
```

```
    #print(sklearn.metrics.accuracy_score(y_train, y_pred))
    k_values.append(i)
    y_pred_val = model.predict(X_val)
    accuracies.append(sklearn.metrics.accuracy_score(y_val,
y_pred_val))
sns.lineplot(x = k_values, y = accuracies)
model = sklearn.neighbors.KNeighborsClassifier(5)
model.fit(X_train, y_train)
y_pred_test = model.predict(X_test)
print(sklearn.metrics.accuracy_score(y_test, y_pred_test))
```

When you run this code, you will see several images, much like what is shown in Figures 13.3 to 13.5.

```
          X      Y
0    1.7500  3.970
1   -0.0547 -2.290
2    3.3500  3.320
3    5.1600  2.930
4    3.0000  1.290
..      ...    ...
495  2.4800  3.110
496  0.3430  0.285
497  3.3900  2.690
498  2.7900  1.920
499  4.4000  2.940
```

FIGURE 13.3 Initial data.

```
[500 rows x 2 columns]
     class
0      0.0
1      1.0
2      1.0
3      0.0
4      0.0
..     ...
495    1.0
496    0.0
497    1.0
498    1.0
499    1.0
```

FIGURE 13.4 KNN two columns.

```
[500 rows x 1 columns]
         X       Y
450   4.790   0.477
451  -0.462  -1.900
452   3.830   2.120
453   0.272   0.919
454   2.600   1.360
455   2.390   2.620
456   4.210   3.260
457   4.040   3.040
458   2.870   3.010
459   1.630   5.400
460  -0.948   0.534
461   5.030   3.910
462  -3.950   2.740
463   0.932  -2.300
464   3.550   2.150
465   5.140   2.050
466   0.496  -1.860
467  -0.238  -0.246
468   4.630   2.650
469   0.962  -0.452
470   4.030   2.870
471   3.830   2.610
472   1.640   2.980
473  -1.560   1.620
474   4.700   3.670
475   0.728  -3.940
```

FIGURE 13.5   KNN one column.

Again, the purpose of this particular algorithm is simply to allow you to see what is happening when you execute the KNN algorithm. Closely examining the two Python scripts produced in this chapter will help you achieve a comfort level with the KNN algorithm.

ALTERNATIVES

There are some well-known alternatives to and variations of k-nearest neighbor. While this chapter's focus is on KNN, it is useful to at least mention some of these alternatives and provide a brief description of them.

One alternative/variation is the nearest centroid classifier. This approach classifies data points based on the class of training samples whose mean is closest to the data point. The mean is the centroid of the class. This approach is sometimes used with text classification in documents. In the case of text classification, it is often called the Rocchio classifier. The algorithm is given a labeled set of training samples $\{\{x_1, y_1\} \ldots (x_n, y_n)\}$ that have labels $l_i \in L$, and which calculate the centroids (represented by μ) with equation 13.4.

$$\mu_L = \frac{1}{1^c L_1} \Sigma_i \in c_L x_i \qquad \text{(eq. 13.4)}$$

Fortunately, the nearest centroid is actually built into scikit-learn, so much of the work is done for you. The following code is a simple example using the breast-cancer dataset.

```
# Importing libraries
from sklearn.neighbors import NearestCentroid
from sklearn.datasets import load _ breast _ cancer
from sklearn.metrics import classification _ report
from sklearn.model _ selection import train _ test _ split
import pandas as pd
# Loading the breastcancer dataset just for an examples
dataset =load _ breast _ cancer()
# From the dataset, seperate data from targets
X = pd.DataFrame(dataset.data)
y = pd.DataFrame(dataset.target)
# Split training and test data
x _ trainset, X _ testset, y _ trainset, y _ testset = train _ test _
split(X, y, test _ size = 0.3, shuffle = True, random _ state = 0)
# Create the Nearest Centroid Classifier
model = NearestCentroid()
# Training the classifier
model.fit(x _ trainset, y _ trainset.values.ravel())
# Display Accuracy on Training and Test sets
print(f"Training Set Score: {model.score(x _ trainset, y _ train-
set) * 100} %")
print(f"Test Set Score: {model.score(X_testset, y_testset) * 100} %")
# Printing classification report of classifier on the test-set data
print(f"Model Classification Report: \n{classification _ report(y _
testset, model.predict(X _ testset))}")
```

When you execute the preceding code, you should see something like what is shown in Figure 13.6.

You can experiment with parameters such as changing the test_size and random_ state when splitting the test and training data to determine what is most effective. This example should be relatively easy for you to execute.

```
E:\Projects\publishing\Machine Learning For Neuroscience>python centroid.py
Training Set Score : 88.69346733668341 %
Test Set Score : 90.05847953216374 %
Model Classification Report :
              precision    recall  f1-score   support

           0       0.96      0.76      0.85        62
           1       0.88      0.98      0.93       109

    accuracy                           0.90       171
   macro avg       0.92      0.87      0.89       171
weighted avg       0.91      0.90      0.90       171
```

FIGURE 13.6 Nearest centroid.

K-d tree is another, quite interesting, alternative. The k-d tree is actually a data structure. The name is an abbreviation of k-dimensional tree. The purpose of the data structure is to organize points in a k-dimensional space. Here is a brief Python example that builds a k-d tree data structure with randomly generated data.

```python
from collections import namedtuple
from operator import itemgetter
from pprint import pformat
#the node class is used to build the tree.
class Node(namedtuple("Node", "location left_child
right_child")):
    def _ _repr_ _(self):
        return pformat(tuple(self))
#this section defines your kd tree datastructure
def kdtree(pointlist, depth: int = 0):
    if not pointlist:
        return None
    k = len(pointlist[0])
    axis = depth % k
    # Sort point list by axis then select the median as pivot
element
    pointlist.sort(key=itemgetter(axis))
    median = len(pointlist) // 2
    # Create node and build the subtrees
    return Node(
        location=pointlist[median],
        left _ child=kdtree(pointlist[:median], depth + 1),
        right _ child=kdtree(pointlist[median + 1 :], depth + 1),
    )
#main function with random data
def main():
    """Example usage"""
    pointlist = [(2, 2), (5, 5), (10, 1), (2, 12)]
    tree = kdtree(pointlist)
    print(tree)
if _ _name_ _ == "_ _main_ _":
    main()
```

When you execute this code it will simply print out the basic data structure as shown in Figure 13.6.

```
E:\Projects\publishing\Machine Learning For Neuroscience>python kdtree.py
((5, 5), ((2, 12), ((2, 2), None, None), None), ((10, 1), None, None))
```

FIGURE 13.7   K-d tree.

There are in fact many variations to the k-d tree itself. There is the implicit k-d tree, min/max k-d tree, quadtree, octree, and many others. These are beyond the scope of this current chapter, but the interested reader can explore those topics after completing this text. Nature[3] has an interesting article discussing many of the k-nearest neighbor variations.

DEEPER WITH SCIKIT-LEARN

You may have noticed that throughout this chapter we have used scikit-learn. Certainly, this library has been used in previous chapters, but in this chapter, every code sample we have has used this library. This is a good time to delve a bit deeper into this important library. You can also refer to the complete reference material on the scikit-learn website.[4]

First, and most obvious, is to examine the sklearn.neighbors module. We have used neighbors.kdtree,. neighbors.NearestNeibhors, and neighbors.NearestCentroid, in this chapter. There are other related algorithms build into scikit-learn. There is neighbors.KernelDensity, which is used for kernel density estimation. There is also neighbors.KNeighbors.Regressor, used for regression with k-nearest neighbors. Another interesting variation is neighbors.RaduiusNeighborsClassifier. In this algorithm, voting is done by neighbors within a given radius as opposed to using metrics such as Euclidean distance or Mahalanobis distance. There are, as of this writing, three variations built into scikit-learn: neighbors. RadiusNeighborsClassifier, neighbors.RadiusNeighbors.Regressor, and neighbors. RadiusNeighborsTransformer.

Next consider the clustering classes found in sklearn.cluster module. There are many, but several implement algorithms we have discussed in previous chapters. Cluster.DBSCAN implements DBSCAN, which we examined in chapter 12. Cluster. kmeans implements the k-means clustering, which we also examined in chapter 12. Cluster.Birch implements the Birch clustering algorithm; this algorithm was not discussed in chapter 12, but is a similar algorithm.

There are other datasets you can work with. In this book you have seen datasets.load_iris in previous chapters, and in this chapter you have used datasets.load_ breastcancer, but others are built in. These include datasets.load_diabetes, datasets. load_digits, and datasets.load_wine. Unfortunately, for our purposes, there are no built-in neuroscience datasets. However, as you have seen previously in this book, loading a dataset is not an overly complicated process.

There are also many other metrics you can use, besides those that have been covered in this book. In this chapter, and others, you have seen sklearn.metrics. accuracy_score. Similar to accuracy_score, you can use sklearn.metrics.average_ precision_score, which, as the name suggests, gets the precision rather than the accuracy. There are many other metrics you can use that are built into scikit-learn,[5] such as the Jaccard similarity coefficient score (scikitlearn. metrics.jaccard_score). The Jaccard similarity score is used to measure the similarity between sample sets. There is also the Cohen's kapa (scikitlearn.metrics.cohen_kappa_score). Cohen's kappa measures inter-rater reliability.

SUMMARY

While this chapter was shorter than most previous chapters, that is because it focused on a single algorithm, k-nearest neighbor. However, this is an important algorithm. It is widely used in classification problems. When dealing with neurological diagnoses, classification is a very common problem. You will find numerous applications for this algorithm when applying machine learning to neuroscience. Readers wishing to delve deeper into k-nearest neighbor may find the following resources useful:

www.ibm.com/topics/knn
www.cs.cornell.edu/courses/cs4780/2018fa/lectures/lecturenote02_kNN.html
www.scholarpedia.org/article/K-nearest_neighbor

EXERCISES

LAB 1: KNN PARKINSON'S DATA

Now we will apply the second algorithm produced in this chapter to a Parkinson's dataset you can download from www.kaggle.com/datasets/vikasukani/parkinsons-disease-data-set (the code is shown here).

```
import numpy as np # linear algebra
import pandas as pd # data processing, CSV file I/O (e.g.,
pd.read _ csv)
import sklearn
import matplotlib.pyplot as plt
import seaborn as sns
from sklearn.utils import shuffle
from sklearn import neighbors, metrics
from IPython.display import display
#import data
df = pd.read _ csv("parkinsons/Parkinssondisease.csv")
#first shuffle data
#then display it
df = shuffle(df, random _ state = 15)
df = df.reset _ index()
X = df.iloc[:, 1:3]
display(X)
y = df.iloc[:, 3:]
display(y)
#now plot the data, before
#working with it
sns.scatterplot(data = df.iloc[:400, :], x = 'X', y = 'Y',
hue = "class")
#now set up data sets
X _ train = X.iloc[:400, :]
X _ val = X.iloc[400:450, :]
X _ test = X.iloc[450:, :]
```

```
y _ train = y.iloc[:400, :]
y _ val = y.iloc[400:450, :]
y _ test = y.iloc[450:, :]
display(X _ test)
#now we can use KNN
model = sklearn.neighbors.KNeighborsClassifier(5)
model.fit(X _ train, y _ train)
y _ pred = model.predict(X _ train)
print(sklearn.metrics.accuracy _ score(y _ train, y _ pred))

y _ pred _ val = model.predict(X _ val)
print(sklearn.metrics.accuracy _ score(y _ val, y _ pred _ val))
plt.figure(figsize=(20,10))
sns.scatterplot(x = X _ train.loc[:, "X"], y = X _ train.loc[:,
"Y"], hue = y _ train.loc[:, "class"], style = y _ pred)
sns.scatterplot(x = X _ val.loc[:, "X"], y = X _ val.loc[:, "Y"],
hue = y _ val.loc[:, "class"], style = y _ pred _ val)
k _ values = []
accuracies = []
for i in range(1, 15):
    model = sklearn.neighbors.KNeighborsClassifier(i)
    model.fit(X _ train, y _ train)
    y _ pred = model.predict(X _ train)
    #print(sklearn.metrics.accuracy _ score(y _ train, y _ pred))
    k _ values.append(i)
    y _ pred _ val = model.predict(X _ val)
    accuracies.append(sklearn.metrics.accuracy _ score(y _ val,
y _ pred _ val))
sns.lineplot(x = k _ values, y = accuracies)
model = sklearn.neighbors.KNeighborsClassifier(5)
model.fit(X _ train, y _ train)
y _ pred _ test = model.predict(X _ test)
print(sklearn.metrics.accuracy _ score(y _ test, y _ pred _ test))
```

LAB 2: KNN VARIATIONS WITH PARKINSON'S DATA

This lab is meant for the more advanced reader. The idea is to load the Parkinson's dataset, as in Lab 1, then, however, implement a variation of KNN for the algorithm. The code given shortly will execute and produce some results. However, you will get an error message. This is provided to give you a start. You will need to find and correct the error.

```
# Importing libraries
from sklearn.neighbors import NearestCentroid
from sklearn.datasets import load _ breast _ cancer
from sklearn.metrics import classification _ report
from sklearn.model _ selection import train _ test _ split
```

```
import pandas as pd
import seaborn as sns
from sklearn.utils import shuffle
from sklearn import neighbors, metrics
from IPython.display import display
#import data
df = pd.read_csv("parkinsons/Parkinssondisease.csv")
#first shuffle data
#then display it
df = shuffle(df, random_state = 15)
df = df.reset_index()
X = df.iloc[:, 1:3]
display(X)
y = df.iloc[:, 3:]
display(y)
#sns.scatterplot(data = df.iloc[:400,:], x = 'X', y = 'Y',
hue = "class")
#now set up data sets
X_train = X.iloc[:400,:]
X_val = X.iloc[400:450,:]
X_test = X.iloc[450:,:]
y_train = y.iloc[:400,:]
y_val = y.iloc[400:450,:]
y_test = y.iloc[450:,:]
display(X_test)
# Create the Nearest Centroid Classifier
model = NearestCentroid()
# Training the classifier
model.fit(X_train, y_train.values.ravel())
# Display Accuracy on Training and Test sets
print(f"Training Set Score: {model.score(x_trainset, y_train-
set) * 100} %")
print(f"Test Set Score: {model.score(X_testset, y_testset) *
100} %")
# Printing classification report of classifier on the test set
set data
print(f"Model Classification Report: \n{classification_report(y_
testset, model.predict(X_testset))}")
```

NOTES

1. www.researchgate.net/publication/232406523_An_Improved_k-Nearest_Neighbor_
   Algorithm_for_Text_Categorization
2. www.statisticshowto.com/mahalanobis-distance/
3. www.nature.com/articles/s41598-022-10358-x
4. https://scikit-learn.org/stable/index.html
5. https://scikit-learn.org/stable/modules/classes.html#sklearn-metrics-metrics

14 Self-Organizing Maps

INTRODUCTION

The self-organizing map (SOM) is an unsupervised machine learning algorithm that produces a low dimensional representation of a higher dimensional data set[1]. SOMs are a variation of artificial neural networks, but they train using competitive learning rather than error-correction learning. Another difference between SOMs and other artificial neural networks is that SOMs do not have an activation function. Instead, the input weights are passed directly to the output.

THE SOM ALGORITHM

We will first present the algorithm as a simple four-step process and explore those steps. Then, we will expand our description of the algorithm with more detail. The concept is to allow you, the reader, to gradually become more comfortable with the algorithm by first encountering a simplified version, then a more detailed version. The basics of the algorithm are given here:

Step1 Initialization: Weights of neurons in the map layer are initialized.

Step 2 Competitive process: A random sample from the dataset is input into the network. The network then calculates the weights of which neurons are most like that input vector. The formula for that is shown in equation 14.1.

$$Distance^2 = \sum_{i=0}^{n}\left(input_i - weight_i\right)^2 \qquad \text{(eq. 14.1)}$$

In equation 14.1, the n is the number of connections. The node with the best result is usually called the best matching unit (or BMU).

Step 3 Cooperative process: Find the proximity neurons of BMU by neighborhood function. This involves calculating the current radius as shown in equation 14.2.

$$\sigma(t) = \sigma_0 e^{\frac{t}{\lambda}} \qquad \text{(eq. 14.2)}$$

In equation 14.2, t is the current iteration of the algorithm, σo is the radius. The lambda symbol, λ, is defined by the formula in equation 14.3.

$$\lambda = \frac{t}{\sigma 0} \qquad \text{(eq. 14.3)}$$

Step 4 Adaptation process: Update the best matching unit and neighbors' weights by moving the values towards the input pattern. Put another way, the weights are

DOI: 10.1201/9781003230588-17

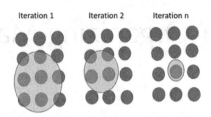

FIGURE 14.1   Self-organizing map.

changed so that the distance becomes smaller. If the maximum count of training iteration is reached, stop. If not, then increment the iteration count by 1 and repeat the process from step 2. Essentially, as the training continues, the radius will get smaller. This process is shown in figure 14.1.

Each neuron on the SOM is assigned a weight vector. This weight vector has the same dimensionality as the input space. The distance between each neuron in the output layer and the input data is computed. The neuron with the least distance is the one that wins the competition. Many methods can be used for distance, but Euclidean distance is commonly used. Euclidean distance was discussed in chapter 13, but is briefly covered again here. Euclidean distance is the length of a line segment between two points. There are many variations of this formula. The first, shown in equation 14.4, is for data that has two dimensions. There is a point p and a point q, with coordinates. The distance between p and q is shown in equation 14.4.

$$d(p,q) = \sqrt{\left(q_1 - p_1\right)^2 + \left(q_2 - p_2\right)^2}$$ (eq. 14.4)

In general, a goal of training an SOM is to represent an n-dimensional input space into a map space with two dimensions. Dimensionality reduction has been discussed extensively in previous chapters. A map consists of nodes/neurons arranged in a rectangular or hexagonal grid with two dimensions. Each neuron/node has a weight vector. The training process leaves the neurons/nodes in place in the map space, but moves the weight vectors towards the input data. When using the aforementioned Euclidian distance, this means reducing the Euclidian distance as much as possible.

The sklearn package has a built-in self-organizing map. The following code is a simple implementation of the self-organizing map.

```
import matplotlib.pyplot as plt
from matplotlib.colors import ListedColormap
from sklearn import datasets
#note this is different than just
#sklearn. You will probably have to use
#pip install sklearn_som
from sklearn_som.som import SOM
# Load iris data. This is a common data set used
# to learn machine learning
```

```
iris = datasets.load _ iris()
iris _ data = iris.data
iris _ label = iris.target
# Extract 2 features
iris _ data = iris _ data[:,:2]
# Build a SOM with 3 clusters
som = SOM(m=3, n=1, dim=2, random _ state=1234)
# Fit it to the data
som.fit(iris _ data)
# Assign each datapoint to its predicted cluster
predictions = som.predict(iris _ data)
# Plot the results
fig, ax = plt.subplots(nrows=2, ncols=1, figsize=(5,7))
x = iris _ data[:,0]
y = iris _ data[:,1]
colors = ['red', 'blue', 'green']
ax[0].scatter(x, y, c=iris _ label, cmap=ListedColormap(colors))
ax[0].title.set _ text('Actual Data')
ax[1].scatter(x, y, c=predictions, cmap=ListedColormap(colors))
ax[1].title.set _ text('What SOM predicted')
plt.savefig('irissom.png')
```

When you run this, you won't see anything in the command window. However, a file named "irissom.png" will be created in the same directory as your Python script. That image is shown in Figure 14.2.

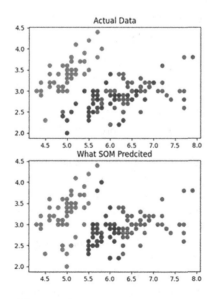

FIGURE 14.2 Self-organizing-map image.

Recall from chapter 13 that there are numerous other data sets and metrics built into scikit-learn. You should consider utilizing some of these to get comfortable with the algorithm.

SOM IN MORE DETAIL

Now that you have seen the SOM algorithm in four steps, and even worked with SOM code, it is time to add more detail. The algorithm in more detail is described here:

Step 1: Randomize the node weight vectors in a map.
Step 2: Randomly pick an input vector D(t), which we will call V(t).
Step 3: Navigate each node in the map.

Step 3a: Use the Euclidean distance formula to determine the similarity between the input vector and the map node's weight vector.
Step 3b: Follow the node that produces the smallest distance. This node will be the best-matching unit or BM.

Step 4: Update the weight vectors of the nodes in the neighborhood of the BMU (including the BMU itself) by pulling them closer to the input vector.

$$Wv(s + 1) = Wv(s) + \theta \ (u, v, s) * \alpha \ (s) * (V(t) - Wv(s))$$

Step 5: Increase s and repeat from step 2 while $s < \lambda \ s < \lambda$

The following list defines the symbols used in the preceding algorithmic steps:

s is the current iteration.
Lambda (λ) is the iteration limit.
t is the index of the target input data vector in the data set V.
V(t) is a target-input data vector.
v is the index of the node in the SOM.
θ (u, v, s) is a restraint due to distance from BMU. This is often called the neighborhood function.
Wv is the current weight vector for node v.
α (s) is a learning restraint due to the progress of iterations.

As with all artificial neural networks that depend on an iterative process, selecting initial weights is crucial. In his original paper, referenced earlier in this chapter, Kohonen advocated a random initiation of weights. However, in recent years the use of principal component analysis for dimensionality reduction to have better weights has been widely used. PCA was introduced in chapter 9 and is briefly described here. Principal component analysis finds lines and planes in the k-dimensional space that approximate the data as closely as possible. Closely is defined using least squares. Least squares is a method to find the line that best fits the data. A line or plane that is the least squares approximation of a set of data points makes the variance of the coordinates on the line or plane as great as possible.

TABLE 14.1
Self-Organizing-Map Parameters

Parameter	Data type	Description
M	Integer	The shape along the vertical dimension of the self-organizing map.
N	Integer	The shape along the horizontal dimension of the self-organizing map.
Lr	Float	The initial step size used to update the self-organzing-map weights.
Dim	Integer	The number of features (i.e., dimensionality) of the input space.
Sigma	Float	This is an optional parameter for the magnitude of changes to each weight.
Max_iter	Integer	Optional parameter for the maximum number of training iterations.
Random_state	Integer	Optional seed for the pseudo random-number generator used for weight initialization.

As you may expect by this point in the book, scikit-learn has self-organizing maps built in. The package is sklearn-som.[2] In fact, you used that package in the previous code example. There is a GitHub repository just for this project.[3] Many parameters for the self-organizing-map package are shown in Table 14.1.

In addition to the previous parameters, there are functions you may wish to be familiar with. The function *fit* simply fits the self-organizing map to the data provided. The function *predict* predicts clusters for each of the elements in the dataset provided. The function *transform* transforms the data into cluster–distance space. The function *fit_predict* essentially calls first fit, then predict, in a single function call. In the code sample earlier in this chapter you saw fit and predict each called separately. You could replace those calls with the single call to fit_predict. The function *fit_transform* works similarly, first calling fit, then transform. You should consider returning to the previous example and experimenting with changing parameters and calling various functions. This will give you an opportunity to become very familiar with self-organizing maps in scikit-learn.

VARIATIONS

The purpose of this chapter is to introduce you to self-organizing maps. We will not be going into details on variations of the SOM, however a brief description of some of the more common variations is given in this section.

GSOM

The growing self-organizing map (GSOM) was created to address the issue of finding a proper map size in the SOM. A GSOM starts with a small number of nodes and then grows new nodes as needed. The algorithm was first published in a 1998 paper.[4]

GSOM sometimes, just called GTOM or even GTM, is a probabilistic algorithm. That means that, unlike SOM, it does not require a shrinking neighborhood, but it is probably convergent. The nodes in a SOM are not constrained, the nodes in a

GSOM/GTOM are constrained. SOM was meant to be a biological model of neurons, whereas GSOM/GTOM is not meant to model actual neurology.

There are a few libraries for generative topographic mapping (another way of writing GSOM) for Python.[5],[6] However, the level of support these libraries have is not clear, so they may not be reliable.

TASOM

Time adaptive self-organizing maps (TASOM) use an adaptive learning rate along with a scaling parameter. For each new input vector, the neighborhood size and the learning rate of the nodes/neurons are updated. There are even variations of the TASOM, such as the binary tree time adaptive self-organizing map (while you may think this should be abbreviated BTTASOM, it is usually just abbreviated as BTASOM).

ELASTIC MAPS

Elastic maps are used for nonlinear dimensionality reduction. The name stems from the algorithm essentially being a construction of elastic springs that are embedded in the dataspace. This particular variation of SOM is important, as it is primarily applied in bioinformatics.

First consider a dataset we will call S that exists in a finite dimensional Euclidean space. The individual datapoints in s are usually simply $s \in S$ (i.e., datapoint s that is an element of the set S). An elastic map is a set of nodes w_j that exist in the same space. Each data point in S has a host node. That host node is simply the closest node w_j to the data point. The dataset is then divided into classes defined as shown in equation 14.5.

$$Kj = \{s \mid wj\} \qquad (eq.\ 14.5)$$

Recall that we stated that these elastic maps are elastic springs. That means they can compress or expand. The stretching energy of these springs is shown in equation 14.6.

$$U_E = \frac{1}{2}\lambda \sum_{(\mathbf{w}_i, \mathbf{w}_j) \in E} \left\| \mathbf{w}_i - \mathbf{w}_j \right\|^2 \qquad (eq.\ 14.6)$$

We already know that the w_i–w_j are nodes. These nodes are connected by elastic edges. The symbols λ and μ are the stretching and bending moduli. The mathematics and details of this algorithm, as you can see, are a bit more complex than others we have examined in this book. And as was stated earlier, this section is just intended to provide a brief introduction to these SOM variations and alternatives. For readers wanting more details on elastic maps, the following resources may be useful.

https://link.springer.com/chapter/10.1007/11941354_120
http://bioinfo-out.curie.fr/projects/elmap/

GROWING SELF-ORGANIZING MAPS

This variation of the self-organizing map is designed to address the issue of finding the proper map size for a self-organizing map. The process of the GSOM is given in the following steps:

1. Initialize the weights of the starting nodes with random numbers between 0 and 1.
2. Calculate the growth threshold (GT) for the data set provided. The dataset is of dimension D with a spread factor of SF, using the formula GT = −D * ln(SF).
3. Input data to the network.
4. Determine the weight vector that is closest to the input vector mapped to the current feature map. Euclidean distance is typically used for this.
5. The weight vector adaptation is applied to the "winner" node and the neighborhood of that node.
6. Increase the error value of the winner.
7. When the total error (TE) of $node_i$ is greater than the growth threshold (GT), if I is a boundary node, then grow nodes. If $node_i$ is not a boundary node, then distribute the weights to the neighbors of $node_i$.
8. If nodes were added in step 7, then initialize the new node weight vectors to match their neighboring nodes.
9. Initialize the learning rate back to its starting value.
10. Repeat.

As you can see, the name of this algorithm stems from the fact that the map can grow new nodes. GSOM is often used in data mining, but can also be used in clustering and classification. There is a GSOM example on GitHub that readers may wish to experiment with.[7]

SUMMARY

This chapter focused on a single algorithm, the SOM. While that makes the chapter shorter than most other chapters in this book, it also lets you focus on mastering a single algorithm. You should take the time to ensure you are fully comfortable with self-organizing maps before continuing past this chapter. Readers wanting to further explore self-organizing maps may find the following resources useful:

www.cs.hmc.edu/~kpang/nn/som.html
https://link.springer.com/book/10.1007/978-3-642-56927-2

EXERCISES

LAB 1: SOM FOR NEUROSCIENCE

In previous chapters the lab had fully working code for you. By now, you should be able to adapt code as needed. This is how you will know you have learned machine learning for neuroscience sufficiently. Therefore, the lab for this chapter takes the code previously discussed, but substitutes in one of the neuroscience datasets introduced in previous chapters. The code provided has intentional blanks you will need to fill in. This is meant to be challenging. After studying 14 chapters and seeing dozens of working code samples, it is now time for you to do some coding without the answer being given to you.

```
import matplotlib.pyplot as plt
from matplotlib.colors import ListedColormap
from sklearn import datasets
#note this is different than just
#sklearn. You will probably have to use
#pip install sklearn _ som
from sklearn _ som.som import SOM
# Load neuroscience data from previous data sets
# discussed in this book.
#*****************you fill this in**********
#you will probably need to adjust the number dimensions, state,
etc.
# Extract 2 features
# Build a SOM with 3 clusters
som = SOM(m=3, n=1, dim=2, random _ state=1234)
# Fit it to the data
som.fit(your _ data)
# Assign each datapoint to its predicted cluster
predictions = som.predict(your _ data)
# Plot the results
fig, ax = plt.subplots(nrows=2, ncols=1, figsize=(5,7))
x = iris _ data[:,0]
y = iris _ data[:,1]
colors = ['red', 'blue', 'green']
ax[0].scatter(x, y, c=iris _ label, cmap=ListedColormap(colors))
ax[0].title.set _ text('Actual Data')
ax[1].scatter(x, y, c=predictions, cmap=ListedColormap(colors))
ax[1].title.set _ text('What SOM predicted')
plt.savefig(yourdata.png')
```

LAB 2: WRITING YOUR OWN CODE

This will be one of the most challenging labs you have had. You will take any of the neuroscience datasets from previous chapters, and make them work with a SOM. There is no code given to start you on this. Here are datasets previously used in this book:

Dementia:
www.kaggle.com/code/dhwanimodi239/demo-dementia-classification/
    data?select=oasis_longitudinal.csv

Alzheimer's:
https://www.kaggle.com/datasets/jboysen/mri-and-alzheimers
EEG Emotions:
www.kaggle.com/datasets/birdy654/eeg-brainwave-dataset-feeling-emotions
Brain-tumor dataset:
www.kaggle.com/datasets/preetviradiya/brian-tumor-dataset

NOTES

1. https://sci2s.ugr.es/keel/pdf/algorithm/articulo/1990-Kohonen-PIEEE.pdf
2. https://pypi.org/project/sklearn-som/
3. https://github.com/rileypsmith/sklearn-som
4. www.microsoft.com/en-us/research/wp-content/uploads/1998/01/bishop-gtm-ncomp-98.pdf
5. https://pypi.org/project/ugtm/
6. https://pypi.org/project/pygtm/
7. https://github.com/philippludwig/pygsom/blob/master/gsom.py

Index

Page numbers in *italics* indicate figures; page numbers in **bold** indicate tables.

Printed in the United States
by Baker & Taylor Publisher Services